Zumwalt-Class
A Reference for the 21st Century Destroyer

Contents

Chapter 1

Overview of Zumwalt-Class Destroyer

1.1 SC-21 (United States)

The distinctive hull of the Zumwalt *class was derived from the SC-21 program.*

SC-21 (Surface Combatant for the 21st century) was a research and development program started in 1994 intended to design land attack ships for the United States Navy. A wide variety of designs were created and extensively examined, including an arsenal ship with 500 cruise missiles. Eventually a "tumblehome" design of around 16,000 tons with two long-range guns and 128 missile tubes was selected as the **DD-21**, the **Destroyer for the 21st century**. The program ended in November 2001, with a version of the DD-21 emerging as the DD(X) or *Zumwalt* class destroyer. It was envisaged that the DD-21 hull would be used for a future air defense cruiser (CG-21), which then eventually evolved into the CG(X) program.

1.1.1 Background

The origins of SC-21 lie in the realization by Admiral Joseph Metcalf III that new technologies such as vertical launch missiles permitted a complete rethink of warship design. He established a steering group, Group Mike, to study the possibilities.[1] Group Mike sponsored two studies in 1987, the Ship Operational Characteristics Study (SOCS) and the Surface Combatant Force Requirement Study (SCFRS).[1] SOCS sought to identify the operational characteristics required of an escort ship, and SCFRS estimated how many such ships were required by the fleet.[1] Since it was expected at that time that the Navy would be fighting prolonged campaigns in the Norwegian Sea, SOCS put an emphasis on ships' continuing ability to fight after an initial Soviet attack.[1] This in turn called for larger, more survivable escort ships than had historically been the norm, around 12,000 tons, and for networking sensors and weapons together so that they could be used by the task force as a whole even if an individual ship had their radar disabled.[1] Survivability also called for the bridge and Combat Information Center to be combined and "buried" in the heart of the ship, and for the ship to use electric drive to distribute the engineering around the ship, which would also give more room for weapons and the scope for railguns and laser weapons in future.[1] SCFRS suggested that the Navy should not replace the *Perry* class frigate for convoy escort duties, but concentrate on building front-line combatants that could be assigned to less demanding convoy duties in their later years.[1]

Both studies reported in 1989, and almost immediately were rendered obsolete by the ending of the Cold War. The Navy suddenly faced the disappearance of its greatest threat, and the prospect of budget cuts as part of the peace dividend. Interest waned in big new designs like the SOCS ship; the Destroyer Variant (DDV) program of December 1991 was intended as a stopgap, the final development of the *Burke* class destroyer.[2]

In 1992 the CNO ordered a 21st-century Destroyer (DD-21) Technology Study.[2] This led to a new program called Surface Combatant for the 21st century (SC-21), intended as a family of ships with a range of capabilities that would not necessarily fit old designations of "destroyers" and "cruisers".[2] Meanwhile strategy papers such as "FORWARD...FROM THE SEA" were redefining the Navy's priorities towards littoral warfare and the support of amphibious assaults inland.[3] It seemed, then, that land attack would be the most important mission for the new ships.[2]

Naval fire support role

Main article: United States Naval Gunfire Support debate

Since the retirement of the Iowa-class battleships, there had been a Congress-mandated requirement relating to the Navy's capability for Naval Fire Support (NFS). The U.S. Marine Corps and the U.S. Navy maintained that destroyers would be adequate in this role, although there are dissenters.[4]

While smaller caliber guns (and missiles) have been used for centuries in naval fire support, very large guns have special capabilities beyond that of mid-range calibres. US battleships were re-activated three times after WWII specifically for NFS, and their 16 inch gunfire was used in every major engagement of the U. S. from WWII to the Gulf War. *Iowa* and *Wisconsin* were finally struck from the Naval Vessel Register in 2006, having been kept on in part to fill a naval fire support role.

1.1.2   Program approved

The SC-21 Mission Need Statement was approved by the Joint Requirements Oversight Council between September–October 1994.[5] The Defense Acquisition Board approved the project on 13 January 1995,[6][7] allowing the program to proceed to Cost & Operational Effectiveness Analysis (COEA).

1.1.3   Concept designs

The SC-21 COEA had an unusually wide remit, and studied a variety of designs from 2,500 tons to 40,000 tons.[8] There were three main "concepts". Concept 1 looked at possible upgrades to existing vessels, Concept 2 looked at variations of existing designs, and Concept 3 was for new ships :[9]

- 2A : newbuilds of *Arleigh Burke* Flight IIA

- 2B : further update of the *Burke* design

- 3A : Power Projection Ship, Aviation Cruiser, Heavy Cruiser – most had 256 VLS cells and amphibious capability

- 3B : Littoral Combatant - Affordable multimission ship with 128 VLS; similar to Improved *Spruances*

- 3C : Maritime Combatant, Armed Supertanker, Agile Maritime Patrol Ship, Small ASW Combatant, Focused Mission Local Area Combatant - 8-64 VLS

- 3D : Expeditionary Force Support Ship, Tailored Maritime Support Ship and other vessels with modular "mission packs".

Option 3B1 was closest to what became the *Zumwalt* class, with a pair of 64-cell VLS fore and aft and two standard 5" guns on a conventional flared hull of around 9,400 tonnes. A bigger hull would be required to enclose everything in a stealthy shape, and to accommodate the much bigger AGS gun system.[10]

1.1.4   Arsenal ship

Main article: Arsenal ship

In a separate study in 1993, two French students had been assigned the design of a Large Capacity Missile Ship, a 20,000-tonner with 500 VLS cells filled with land-attack missiles.[11] This design was inspired by a RAND paper in that year, which suggested a land invasion could be halted by destroying 20% of its vehicles with precision munitions.[12] This would take several days with aircraft, but a surface ship with large numbers of land-attack missiles could achieve the same effect almost instantly.

This design was apparently included in the SC-21 assessment as an afterthought - it was not included in the original list of concepts.[11] Two designs were considered, both with 512 VLS cells - Option 3A6 was a minimal version of 13,400 tons with no self-defense capability, Option 3A5[13] was a 30,000 ton "goal ship" with many more survivability features.[11] The latter Maritime Fire Support Ship became the basis of the Arsenal Ship championed by CNO Jeremy Boorda. In fact he was so enthusiastic that the rest of the SC-21 program was suspended in favor of development of the Arsenal Ship.[11]

The Navy set up a joint venture with DARPA on March 18, 1996.[14] The Arsenal Ship would be acquired as a prototype under DARPA's Other Transaction Authority under Section 845 of the National Defense Authorization Act for FY 1994 (Public Law 103-160),[15] which allowed them to bypass much of the bureaucracy involved with defense procurement, enabling a prototype to be built by the end of 2000. The requirement was for a network-capable ship with around 500 VLS and less than 50 personnel, for a cost of less than $520 m for the lead ship.[14] A further five ships would be acquired at a later date.

In July 1996, five consortia were awarded $1 m to come up with some concepts.[16] Three received follow-on contracts in January 1997, but the Navy had lost enthusiasm for the project with Boorda's suicide in May 1996, and in April 1997 the Arsenal Ship was redesignated as the Maritime Fire Support Demonstrator (MFSD), which would be a technology demonstrator for a revitalised SC-21 program.[16] As a result, Congress cut funding to the project and it was finally canceled in October 1997.[16] The Arsenal Ship concept was revived in 2002 by converting four *Ohio*-class submarines into SSGNs carrying 154 VLS tubes.

1.1.5 DD-21

In 1997, plans for the Littoral Combatant (3B1) were revived under the SC-21 banner. It was initially renamed the Power Projection Ship, and then DD-21, Destroyer for the 21st century. Influenced by the Arsenal Ship, it would have a stealthy hull with a significant land attack capability. At first the plan was to install a twin-barreled Vertical Gun for Advanced Ships (VGAS), developed from experiments on advanced projectiles for the *Iowa* battleships, but this was dropped in favour of a conventional 5" gun and two 64-cell VLS. It would also feature a revolutionary cross-layer active sonar.[17] An Operational Requirements Document was signed in November 1997,[18] and an Advanced Development Memorandum on 11 December.[17] A Program Executive Office was established on 25 February 1998.[17] As with the *Burke* class, the construction of the DD-21 ships would be split between Bath Iron Works and Ingalls to preserve the industrial base. However there would be a competition between the two yards to design DD-21 and to be the full-service contractor for the class, which would mean the winning team receiving 85% of the total program costs of around $70 billion.[19] BIW partnered with Lockheed Martin as the combat system designer and integrator, forming the "Blue" team; Ingalls partnered Raytheon in the "Gold" team.[20]

The Zumwalt *class tumblehome hull is derived from that of the DD-21*

The new design was compared to the current *Burke* hull and possible developments of it, and it was decided to go ahead with a new hull[17] to be called **DD-21** (or **DD 21**),[21] the **21st century Land Attack Destroyer**.[22] The new hull

was judged to have more potential for stealth and reductions in manning than the *Burkes*. This was important, as one aim of the program was to reduce manning and operational costs by 70%, while providing scope for a follow-on cruiser class.[23] On 4 July 2000 it was announced that the lead ship of the class would be named after Admiral Elmo Zumwalt, who had died earlier that year.[24] The name would be inherited by subsequent versions of the design, culminating in USS *Zumwalt*.

The ship's delivery schedule was delayed by a year[25] following the decision in January 2000 to use electric drive in the ship.[26] But by early 2000 a "tumblehome" design had emerged that resembles that of the eventual *Zumwalt* class, albeit with differences in the weapons carried.[27] Sources disagree on the displacement of the DD-21, and indeed it probably varied during the design process, but around 16,000 tons seems most likely.[28] As of July 2001, 32 DD-21s were planned, with construction planned to start in FY05.[29]

Many of the weapons planned for the DD-21 were to be trialled in existing ships, increasing the land attack capability of the existing fleet in FY05 before the delivery of DD-21 in FY10.[30] Rocket-boosted Extended Range Guided Munitions (ERGM) for existing 5"/62 guns would have a range of 63 nautical miles (117 km), while the long-range Block IV Tactical Tomahawk missile could be fired from existing Mk 41 vertical launching systems.[31] It was also planned to turn old Standard Missile SAMs into Land Attack Standard Missiles (LASM) with 150 nautical miles (278 km) range.[32]

Initially it was planned to use the Vertical Gun for Advanced Ships (VGAS),[33] but this was abandoned in favor of a more conventional Advanced Gun System fore and aft, each with a separate magazine of 600-750 rounds.[34] The guns would fire a 155 mm version of the ERGM which would double the payload and increase the range to 100 nautical miles (185 km).[35] Together the two guns[29] would give the ship a rate of fire of 24 rounds/minute, giving them the throw weight of two 6-gun 155 mm artillery batteries.[36] Precision munitions make gunfire three times more effective than unguided shells, hence the DD-21 was said to have the destructive power of six batteries.[36] The Navy's goal was to have 256 VLS cells on the DD-21,[37] but the final number may have been 128[38] - some sketches show missiles being launched from fore and aft[39] but the aft launchers appear to have been replaced by a second gun system and/or a helicopter pad. As well as LASM and Tactical Tomahawk, the DD-21 would receive the Advanced Land Attack Missile (ALAM), a new missile with a variety of warheads and a design range of up to 300 nautical miles (556 km).[40]

1.1.6 CG-21

A **21st century air defense cruiser** (CG-21) was announced in January 2000 to replace the 27 *Ticonderoga* class cruisers.[41] Procurement was to begin after the end of the DD-21 program, perhaps around 2015.[38] Development work had not started before the program was terminated in November 2001;[42] CG-21 was replaced by the CG(X) program, which was subsequently cancelled in 2010.

1.1.7 Cancellation

The winner of the competition to design the DD-21 was due to be announced in March 2001, but the decision was put back twice as the new Bush administration reviewed defense spending. On March 1 it was announced that the decision would be made in May, and on May 31 it was announced that the Navy would wait for the results of the Quadrennial Defense Review, and a future shipbuilding review.[43] After the House Appropriations Committee proposed a reduction in the DD-21 allocation in the FY2002 budget in late October 2001,[44] on 1 November the Navy announced a less ambitious Future Surface Combatant program (FSC). Polmar claims that DD-21 was terminated primarily for political reasons as the program was closely identified with the Clinton administration,[45] whereas Work views it as the culmination of a debate within the Navy about whether they should use in the littoral zone large capable ships like the DD-21 or more numerous smaller ships like the "Streetfighter" concept.[46] It did not help that the original plan called for the fifth ship to cost $750 m in FY96 dollars,[23] but in the fourth quarter of 1999 alone the program cost went up from $3.2bn to $5.2bn.[47]

Streetfighter evolved into the Littoral Combat Ship; under FSC the DD-21 became the DD(X) which would become the *Zumwalt* class destroyer, whilst the preliminary plans for CG-21 would be folded into the CG(X) ballistic missile defense cruiser.[16]

The hull of the *Zumwalt* class is similar to that of the DD-21, but the new design displaces 14,564 tons[48] and unlike the DD-21, the deckhouse is flush to the sides of the hull. The central "block" of VLS cells is replaced by a peripheral VLS of 80 cells,[48] which allows both guns to be located forward of the deckhouse. This in turn allows the stern to be given over to helicopter facilities but means that the automated magazine can only contain 750 rounds,[49] supplemented by an auxiliary store.

1.1.8 References

Notes

[1] Friedman, Norman (2004), *U.S. Destroyers: An Illustrated Design History*, Naval Institute Press, pp. 431–3, ISBN 978-1-55750-442-5

[2] Friedman, pp434-5

[3] *FORWARD...FROM THE SEA* (PDF), Department of the Navy, 1994

[4] Novak, Robert (2005-12-06), *Losing the Battleships*, CNN

[5] *JROC To Weigh Anchor On New USN Warship*, archived from the original on February 27, 2009) said that a draft of the MNS had been circulating since late 1993, but JROC had sent it back to the Navy in June to expand its scope to options other than destroyer/cruiser class ships. It was not until 15 October that *Jane's* reported final approval.

[6] *Program Executive Office Ships - DDG1000 Program History*, US Navy, 2008-10-09, retrieved 2008-10-18

[7] Kaminski, Paul G. (1995-01-18), *Surface Combatant 21 (SC-21) Acquisition Decision Memorandum (ADM)* Copy on FAS website

[8] "JROC Head Approves SC-21 MNS", *Jane's Defence Weekly*, 1994-10-15, archived from the original on February 27, 2009

[9] Friedman pp432-440 has more details and sketches of several of the SC-21 concept ships.

[10] Friedman, p436

[11] Friedman, p440-3

[12] Bowie, Christopher J; et al. (1993), *The New calculus: analyzing airpower's changing role in joint theater campaigns*, RAND Corporation, ISBN 0-8330-1322-X, MR-149-AF

[13] 3A5 or 3B5? Friedman seems confused

[14] Leonard, Robert S.; Drezner, Jeffrey A.; Sommer, Geoffrey, *The Arsenal Ship Acquisition Process Experience Contrasting and Common Impressions from the Contractor Teams and Joint Program Office*, ISBN 0-8330-2690-9 Appendix F, Joint Memorandum—Arsenal Ship Program - the rest of the report is a good history of the Arsenal Ship program.

[15] *Section 845 Other Transactions For Prototypes* (PDF), Contracts Management Office, DARPA, retrieved 2008-10-18

[16] Van Atta, Richard H.; Lippitz, Michael J.; Lupo, Jasper C.; Mahoney, Rob; Nunn, Jack H. (April 2003), *Transformation and Transition: DARPA's Role in Fostering an Emerging Revolution in Military Affairs Volume 1 – Overall Assessment* (PDF), Institute of Defense Analyses, pp. 32–33

[17] Friedman p445

[18] Friedman p445 says the ORD was signed in November 1997; Slide 9 of the Heller briefing says the ORD was dated 9 October 1997.

[19] *Navy Puts $30 Billion DD 21 Design Contract on Hold*, Washington: Reuters (republished by Marinelink.com), 2001-06-01

[20] O'Rourke, Ronald (2004-10-24), *Navy DD(X) and LCS Ship Acquisition Programs: Oversight Issues and Options for Congress*, Congressional Research Service, RL32109 page CRS-11

[21] The hyphenated "DD-21" is most commonly used, by analogy with SC-21 and standard hull numbers, but the "house style" of different departments may vary. For instance the DoD never uses hyphens in the hull numbers of ships on www.defenselink.mil, so the Ticonderoga cruisers are referred to as CG 47 on that site, and DD-21 as DD 21. "DD 21" (but never "DD21") seems to have become more common after the end of SC-21, for instance the Heller briefing of June 2000 uses DD 21.

[22] Polmar, Norman (2004), *The Naval Institute Guide to the Ships and Aircraft of the U.S. Fleet*, Naval Institute Press, ISBN 978-1-59114-685-8

[23] Heller, Jim (2000-06-21), *DD 21 Advanced Missile and Gun Systems* (PDF), Defense Technical Information Center, p. 4 Briefing to National Defense Industrial Association by the DD 21 deputy project manager.

[24] *President Names New Ship Class After Admiral Zumwalt*, U.S. Department of Defense, 2000-07-04

[25] Polmar, Norman (2001), *The Naval Institute Guide to the Ships and Aircraft of the U.S. Fleet*, Naval Institute Press, p. 142, ISBN 978-1-55750-656-6

[26] *Integrated Power Systems, Electric Drive Selected For New Navy Destroyers (DD 21)*, U.S. Department of Defense, 2000-01-06

[27] For the shape of the DD-21 in early 2000 see the pictures in the Heller and Hamilton briefings. Although the text and other pictures make clear that two guns were planned for the ship, some sketches show only one gun. These may predate the decision to use electric drive which would have freed up space for a second gun in the stern.

[28] O'Rourke (2004) page CRS-12 says that the DD-21 was reported as 16,000 tons, then cites John Young talking about plans for 17,000 tons. Polmar (2004) p146 says "15,000-17000 tons" was most likely.

[29] *10-K for 12/31/00*, United Defense Industries Inc, 2001-07-03 SEC Files 333-43619, −01, −02, −03; Accession Number 1021408-1-1525

[30] Hamilton, Charles (2000-06-20), *Naval Surface Fire Support* (PDF), Defense Technical Information Center, p. 15 Briefing to National Defense Industrial Association by Program Executive Officer Surface Strike.

[31] Hamilton pp4-6

[32] Hamilton p21

[33] Polmar (2001) p126; p481 goes into more detail on the VGAS

[34] Heller pp14-15

[35] Hamilton p13

[36] Hamilton p26

[37] Polmar (2001) p143

[38] O'Rourke (2004) page CRS-10

[39] Hamilton p25 shows missiles launching from fore and aft, Heller p15 may show missile cells either side of the aft gun, but Heller p10 appears to show a helicopter pad

[40] Hamilton p25

[41] Polmar (2004) p138

[42] O'Rourke, Ronald (2008-08-08), *Navy DDG-1000 and DDG-51 Destroyer Programs: Background, Oversight Issues, and Options for Congress* (PDF), Congressional Research Service, p. 31

[43] *Navy Delays DD 21 Source Selection Decision*, U.S. Department of Defense, 2001-05-31

[44] O'Rourke (2004) page CRS-13

[45] Polmar (2004) p133

[46] Work, Robert O. (February 2004), *Naval Transformation and the Littoral Combat Ship* (PDF), Center for Strategic and Budgetary Assessments, pp. 45–60 Good history of the debate over DD-21 versus Streefighter, and some of the thinking behind the DD-21 design

[47] *Selected Acquisition Reports From December 1999*, U.S. Department of Defense, 2000-04-13

[48] *DDG 1000 Flight I Design*, Northrop Grumman Ship Systems, 2007

[49] *Advanced Gun System (AGS)*, BAe Systems, 2008, ISBN 1-4235-4061-1

Bibliography

- O'Rourke, Ronald, *Navy Zumwalt (DD-21) Class Destroyer Program: Background and Issues for Congress*, Congressional Research Service - CRS Report RS21059, from around the time of cancellation, is a good review of DD-21 if you can find it.

- *Surface Warfare Magazine* **25** (3), March 2000, ISSN 0145-1073 Missing or empty |title= (help) has a number of articles about the DD-21

- Driesbach, Dawn H. (December 1996), *The Arsenal Ship and the U.S. Navy: A Revolution in Military Affairs Perspective*, Naval Postgraduate School, Monterey (HTML version available from FAS) Overview of the Arsenal Ship.

1.1.9   External links

- SC-21 at Global Security

1.2   Zumwalt-class destroyer

The *Zumwalt*-class destroyers are a class of United States Navy guided missile destroyers designed as multi-mission stealth ships with a focus on land attack. The class emerged from the previous DD-21 vessel program. The program was previously known as the "DD(X)". The class is multi-role and designed for surface warfare, anti-aircraft warfare, and naval gunfire support. They take the place of battleships in filling the former congressional mandate for naval fire support,[9] though the requirement was reduced to allow them to fill this role. The vessels' appearance has been compared to that of the historic ironclad warship.[10]

The class has a low radar profile; an integrated power system, which can send electricity to the electric drive motors or weapons, the Total Ship Computing Environment Infrastructure (TSCEI),[11] automated fire-fighting systems, automated piping rupture isolation, and may someday include a railgun[12] or free-electron lasers.[13] The class is designed to require a smaller crew and be less expensive to operate than comparable warships. It has a wave-piercing tumblehome hull form whose sides slope inward above the waterline. This will reduce the radar cross-section, returning much less energy than a conventional flare hull form.

The lead ship is named *Zumwalt* for Admiral Elmo Zumwalt, and carries the hull number DDG-1000. Originally 32 ships were planned, with the $9.6 billion research and development costs spread across the class, but the quantity was reduced to 24, then to 7, and finally to 3, greatly increasing the cost-per-ship.[14][15] The cost increase caused the U.S. Navy to identify the program as being in breach of the Nunn–McCurdy Amendment on 1 February 2010.[16][17]

1.2.1   History

Background and funding

Many of the features were developed under the DD21 program ("21st Century Destroyer"), which was originally designed around the Vertical Gun for Advanced Ships (VGAS). In 2001, Congress cut the DD-21 program by half as part of the SC21 program; to save it, the acquisition program was renamed as DD(X) and heavily reworked.

Originally, the Navy had hoped to build 32 destroyers. That number was reduced to 24, then to 7, due to the high cost of new and experimental technologies.[14] On 23 November

2005, the Defense Acquisition Board approved a plan for simultaneous construction of the first two ships at Northrop's Ingalls yard in Pascagoula, Mississippi and General Dynamics' Bath Iron Works in Bath, Maine. However, at that date, funding had yet to be authorized by Congress.

In late December 2005, the House and Senate agreed to continue funding the program. The U.S. House of Representatives allotted the Navy only enough money to begin construction on one destroyer, as a "technology demonstrator". The initial funding allocation was included in the National Defense Authorization Act of 2007.[14] However, this was increased to two ships by the 2007 appropriations bill[18] approved in September 2006, which allotted US$2,568m to the DDG-1000 program.[19]

On 31 July 2008, U.S. Navy acquisition officials told Congress that the service needed to purchase more *Arleigh Burke*-class destroyers, and no longer needs the next-generation DDG-1000 class,[20] Only the two approved destroyers would be built. The Navy said the world threat picture had changed in such a way that it now makes more sense to build at least eight more *Burke*s, rather than DDG-1000s.[20] The Navy concluded from fifteen classified intelligence reports that the DDG-1000s would be vulnerable to forms of missile attacks.[21] Many Congressional subcommittee members questioned that the Navy completed such a sweeping re-evaluation of the world threat picture in just a few weeks, after spending some 13 years and $10 billion on development of the surface ship program known as DD-21, then DD(X) and finally, DDG-1000.[20] Subsequently, Chief of Naval Operations Gary Roughead cited the need to provide area air defense and specific new threats such as ballistic missiles and the possession of anti-ship missiles by groups such as Hezbollah.[22] The mooted structural problems have not been discussed in public. Navy Secretary Donald Winter said on 4 September that "Making certain that we have – I'll just say, a destroyer – in the '09 budget is more important than whether that's a DDG 1000 or a DDG 51".[23]

On 19 August 2008, Secretary Winter was reported as saying that a third *Zumwalt* would be built at Bath Iron Works, citing concerns about maintaining shipbuilding capacity.[24] House Defense Appropriations Subcommittee Chairman John Murtha said on 23 September 2008 that he had agreed to partial funding of the third DDG-1000 in the 2009 Defense authorization bill.[25]

A 26 January 2009 memo from John Young, the US DoD's top acquisition official, stated that the per ship price for the *Zumwalt*-class destroyers had reached $5.964 billion, 81 percent over the Navy's original estimate used in proposing the program. If true, that means that the program has breached the Nunn–McCurdy Amendment, requiring the Navy to re-certify and re-justify the program to

Congress.[26]

On 6 April 2009, Defense Secretary Robert Gates announced that DoD's proposed 2010 budget will end the DDG-1000 program at a maximum of three ships.[27] Also in April, the Pentagon awarded a fixed-price contract with General Dynamics to build the three destroyers, replacing a cost-plus-fee contract that had been awarded to Northrop Grumman. At that time, the first DDG-1000 destroyer was expected to cost $3.5 billion, the second approximately $2.5 billion, and the third even less.[28]

What had once been seen as the backbone of the Navy's future surface fleet[29] with a planned production run of 32, has since been replaced by destroyer production reverting to the *Arleigh Burke* class after ordering three *Zumwalt*s.[30] The Zumwalt's failure to meet cost and schedule estimates[31] is a result of a positive feedback loop of spiraling costs and plummeting productions numbers described by Chuck Spinney as the "death spiral", joining other projects such as the F-22, F-35, and Future Combat System.[32]

Construction

Representatives from Naval Sea Systems Command and Bath Iron Works sign a construction contract at the Pentagon, February 2008.

In late 2005, the program entered the detailed design and integration phase, for which Raytheon is the Mission Systems Integrator. Both Northrop Grumman Ship Systems and General Dynamics Bath Iron Works share dual-lead for the hull, mechanical, and electrical detailed design. BAE Systems Inc. has the advanced gun system and the MK57 VLS. Almost every major defense contractor (including Lockheed Martin, Northrop Grumman Sperry Marine, L-3 Communications) and subcontractors from nearly every state in the U.S. are involved to some extent in this project, which is the largest single line item in the Navy's budget. During the previous contract, development and testing of 11 Engineering Development Models (EDMs) took place:

Advanced Gun System, Autonomic Fire Suppression System, Dual Band Radar [X-band and L-band], Infrared, Integrated Deckhouse & Apertures, Integrated Power System, Integrated Undersea Warfare, Peripheral Vertical Launch System, Total Ship Computing Environment Infrastructure (TSCEI), Tumblehome Hull Form.

The decision in September 2006 to fund two ships meant that one could be built by the Bath Iron Works in Maine and one by Northrop Grumman's Ingalls Shipbuilding in Mississippi.[18]

Northrop Grumman was awarded a $90M contract modification for materials and production planning on 13 November 2007.[33] On 14 February 2008, Bath Iron Works was awarded a contract for the construction of the USS *Zumwalt* (DDG-1000), and Northrop Grumman Shipbuilding was awarded a contract for the construction of USS *Michael Monsoor* (DDG-1001), at a cost of $1.4 billion each.[34]

Deckhouse of USS Zumwalt *being installed in December 2012*

On 11 February 2009, full-rate production officially began on the first *Zumwalt*-class destroyer.[35] Construction on the second ship of the class, *Michael Monsoor*, began in March 2010.[36] The keel for the first *Zumwalt*-class destroyer was laid on 17 November 2011.[36] This first vessel was launched from the shipyard at Bath, Maine on 29 October 2013.[37]

The construction timetable in July 2008 was:[38]

- October 2008: DDG-1000 starts construction at Bath Iron Works[39][40][41]

- September 2009: DDG-1001 starts construction at Bath Iron Works.[42]

- April 2012: DDG-1002 starts construction at Bath Iron Works[43]

- April 2013: DDG-1000 initial delivery

- May 2014: DDG-1001 delivery

- March 2015: Initial operating capability

- Fiscal 2018: DDG-1002 delivery

The Navy plans for the USS *Zumwalt* to reach initial operating capability (IOC) in 2016. The second ship, the USS *Michael Monsoor*, is to reach IOC in 2018, and the third ship, the USS *Lyndon B. Johnson* (DDG-1002), is to reach IOC in 2021.[44]

1.2.2 Names and hull numbers

In April 2006, the navy announced plans to name the first ship of the class *Zumwalt* after former Chief of Naval Operations Admiral Elmo R. "Bud" Zumwalt Jr.[38] Its hull number will be DDG-1000, which abandons the guided missile destroyer sequence used by the *Arleigh Burke*-class destroyers (DDG-51–), and continues the previous "gun destroyer" sequence from the last of *Spruance*-class, USS *Hayler* (DD-997).

DDG-1001 will be named for Master-at-Arms 2nd Class Michael A. Monsoor, the second Navy SEAL to receive the Medal of Honor in the Global War on Terror (GWOT), the navy announced on 29 October 2008.[45]

On 16 April 2012, the Secretary of the Navy Ray Mabus announced that DDG-1002 will be named for former naval officer and U.S. President, Lyndon B. Johnson.[46]

1.2.3 Design elements

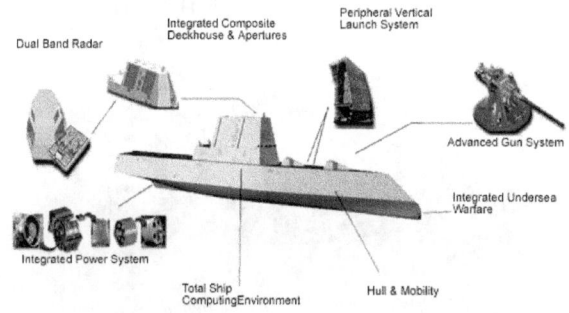

Planned features of the DDG-1000

As of January 2009, the Government Accountability Office (GAO) found that four out of 12 of the critical technologies in the ship's design were fully mature. Six of the critical technologies were "approaching maturity", but five of those would not be fully mature until after installation.[47]

Stealth

Main article: Stealth ship

Despite being 40% larger than an *Arleigh Burke*-class destroyer, the radar cross-section is more akin to that of a fishing boat, according to a spokesman for Naval Sea Systems Command.[48] The tumblehome hull and composite deckhouse reduces radar return. Overall, the destroyer's angular build makes it "50 times harder to spot on radar than an ordinary destroyer.[48]

Zumwalt's deckhouse in transit on 6 November 2012

The acoustic signature is comparable to that of the *Los Angeles*-class submarines. Water sleeting along the sides, along with passive cool air induction in the mack reduces infrared signature.

The composite deckhouse encloses much of the sensors and electronics.[49] In 2008, Defense News reported there had been problems sealing the composite construction panels of this area; Northrop Grumman denied this.[50]

The U.S. Navy solicited bids for a lower cost steel deckhouse as an option for DDG-1002, the last *Zumwalt* destroyer, in January 2013.[51][52][53] On 2 August 2013, the US Navy announced it was awarding a $212 million contract to General Dynamics Bath Iron Works to build a steel deckhouse for destroyer *Lyndon B. Johnson* (DDG-1002).[53]

Tumblehome wave piercing hull

Main articles: Tumblehome and Wave-piercing hull

The *Zumwalt*-class destroyer reintroduces the tumblehome hull form, a hull form not seen to this extent since the Russo-Japanese War in 1905. Originally put forth in modern steel battleship designs by the French shipyard Forges et Chantiers de la Méditerranée in La Seyne in Toulon, French naval architects believed that tumblehome, in which the beam of the vessel narrowed from the water-line to the upper deck, would create better freeboard, greater seaworthiness, and, as Russian battleships were to find, would be ideal for navigating through narrow constraints (e.g. canals).[54] On the down side, the tumblehome battleships experienced losses in watertight integrity and/or stability problems (especially in high speed turns).[55] 21st century tumblehome is being reintroduced to reduce the radar return of the hull. The inverted bow is designed to cut through waves rather than ride over them.[56][57] The stability of this hull form in high sea states has caused debate among naval architects. The tumblehome has not been featured in any USN concept design since the *Zumwalt*-class.

Advanced Gun System

Main article: Advanced Gun System

There has been research on extending the range of naval gunfire for many years. Canadian engineer Gerald Bull and Naval Ordnance Station Indian Head tested an 11 inch (279 mm) sub-caliber saboted long-range round[58] in a stretched 16-inch/45 (406 mm) Mark 6 battleship gun in 1967.[59] The Advanced Gun Weapon System Technology Program (AGWSTP) evaluated a similar projectile with longer range in the 1980s.[58] After the battleships were decommissioned in 1992, the AGWSTP became a 5-inch (127 mm) gun with an intended range of 180 kilometers (110 mi), which then led to the Vertical Gun for Advanced Ships (VGAS). The original DD-21 was designed around this "vertical gun", but the project ran into serious technology/cost problems and was radically scaled back to a more conventional 6.1 inch (155 mm) Advanced Gun System (AGS). One advantage of this move was that the gun was no longer restricted to guided munitions.

The Advanced Gun System is a 155 mm naval gun, two of which would be installed in each ship. This system consists of an advanced 155 mm gun and the Long Range Land Attack Projectile.[60] This projectile is a rocket with a warhead fired from the AGS gun; the warhead weighs 11 kg / 24 lb and has a circular error of probability of 50 meters. This weapon system will have a range of 83 nautical miles (154 km);[Note 1][48] the fully automated storage system will have room for up to 750 rounds.[56][60] The barrel is water-cooled to prevent overheating and allows a rate of fire of 10 rounds per minute per gun. The combined firepower from a pair of turrets gives each *Zumwalt*-class destroyer firepower equivalent to 12 conventional M198 field guns.[61]

In order to provide sufficient stability to fire these guns, the *Zumwalt* will use ballast tanks to lower itself into the water.[62]

Peripheral Vertical Launch System

Main article: Vertical launching system

The Peripheral Vertical Launch System (PVLS) is an attempt to reclaim the prized center space of the hull while increasing the safety of the ship from the loss of the entire missile battery and the loss of the ship in the case of a magazine explosion. The system scatters pods of VLS around the outer shell of the ship having a thin steel outer shell and a thick inner shell. The design of the PVLS would direct the force of the explosion outward rather than ripping the ship in half. Additionally, this design keeps the loss of missile capacity down to just the pod being hit.[56][63]

Aircraft and boat features

Two spots will be available on a large aviation deck with a hangar capable of housing two full size SH-60 helicopters.[64] Boat handling is to be dealt within a stern mounted boat hangar with ramp. The boat hangar's stern location meets high sea state requirements for boat operations.[56]

Radar

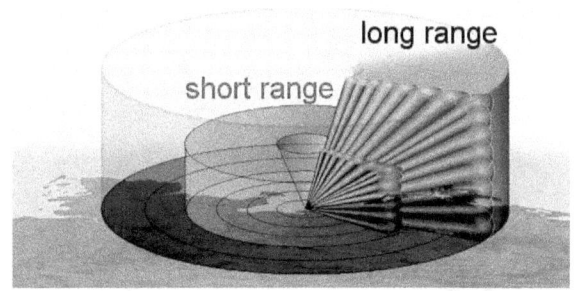

Diagram of AN/SPY-3 vertical electronic pencil beam radar conex projections

Originally, the AN/SPY-3 active electronically scanned array primarily X band radar was to be married with Lockheed Martin's AN/SPY-4 S band volume search radar. Raytheon's X-band, active-array SPY-3 Multi-Function Radar (MFR) offers superior medium to high altitude performance over other radar bands, and its pencil beams give it an excellent ability to focus in on targets. SPY-3 will be the primary radar used for missile engagements.[65] A 2005 report by Congress' investigative arm, the Government Accountability Office (GAO), questioned that the technology leap for the Dual Band Radar would be too much.[5]

On 2 June 2010, Pentagon acquisition chief Ashton Carter announced that they will be removing the SPY-4 S-band Volume Search Radar from the DDG-1000's dual-band radar to reduce costs as part of the Nunn–McCurdy certification process.[30] Due to the SPY-4 removal, the SPY-3 radar is to have software modifications so as to perform a volume search functionality. Shipboard operators will be able to optimize the SPY-3 for either horizon search or volume search. While optimized for volume search, the horizon search capability is limited. The DDG-1000 is still expected to perform local area air defense.[30][66] This system is thought to provide high detection and excellent anti-jamming capabilities particularly when used in conjunction with the Cooperative Engagement Capability (CEC). It is, however, not reported if the CEC system will be installed on the Zumwalt-class destroyers upon commissioning, but it is scheduled for eventual incorporation in the ship type.[67][68]

The Dual Band Radar in its entirety (SPY-3 & SPY-4) is to be installed only on the USS Gerald R. Ford (CVN-78). With the development of the AMDR (Air and Missile Defense Radar), it seems unlikely the DBR is to be installed on any other platforms, as it is on the DDG-1000 class, or in total, as it is on CVN-78. The Enterprise Air Surveillance Radar (EASR) is a new design surveillance radar that is to be installed in the second Ford-class aircraft carrier, John F. Kennedy (CVN-79), in lieu of the Dual Band radar. The America-class amphibious assault ship ships starting with LHA-8 and the planned LX(R) will also have this radar.[69]

AMDR (Air and Missile Defense Radar) was originally proposed to be installed in the hull of DDG-1000 type under the CG(X) program. However, due to cost growth, the CG(X) program was canceled. The AMDR has continued in fully funded development for installation on the DDG-51 Flight III ships. However, a smaller than optimal planned aperture of 14 feet (4.3 m), the AMDR for the Flight III ships is to be less sensitive than the 22 ft variant that had been planned for CG(X).

A study to place the AMDR on a DDG-1000 hull was done with the 22-foot (6.7 m) aperture primarily for Ballistic Missile Defense (BMD) purposes. In that the DDG-1000 does not have an Aegis combat system, as does the DDG-51 class ships, but rather the Total Ship Computing Environment Infrastructure (TSCEI), the Radar/Hull Study stated:

> ... that developing a BMD capability "from scratch" for TSCE was not considered viable enough by the study team to warrant further analysis, particularly because of the investment already made in the Aegis program. The navy concluded that developing IAMD software and hardware specifically for TSCE would be more expensive and present higher risk. Ultimately, the navy determined that Aegis was its preferred

combat system option. Navy officials stated that Aegis had proven some BMD capability and was widely used across the fleet, and that the navy wanted to leverage the investments it had made over the years in this combat system, especially in its current development of a version that provides a new, limited IAMD capability.[70]

Common Display System

The ship's Common Display System is nicknamed "keds": Sailors operate keds via "trackballs and specialized button panels," with the option to "interface by using touchscreens". The technology array allows sailors to monitor multiple weapons systems or sensors, saving manpower, and allowing it to be steered from the ops center.[48]

Sonar

A dual-band sonar controlled by a highly automated computer system will be used to detect mines and submarines. It is claimed that it is superior to the Burke's sonar in littoral ASW, but less effective in blue water/deep sea areas.[71]

- Hull-mounted mid-frequency sonar (AN/SQS-60)

- Hull-mounted high-frequency sonar (AN/SQS-61)

- Multi-function towed array sonar and handling system (AN/SQR-20)[72]

Although Zumwalt ships have an integrated suite of under-sea sensors and a multi-function towed array, they are not equipped with onboard torpedo tubes, so they rely on their helicopters or ASROC missiles to destroy submarines that the sonar picks up.[62]

Propulsion and power system

Main article: Permanent-magnet electric motor §
Permanent-magnet motors

The DDX proposed to use a permanent-magnet motor (PMM) within the hull. An alternate twin pod arrangement was rejected as the ramifications of pod drives would require too much development and validation cost to the vessel. The PMM is considered to be another technology leap and is the cause of some concern (along with the radar system) from Congress.[56] As part of the design phase, Northrop Grumman had built the world's largest permanent magnet motor, designed and fabricated by DRS Technologies. This proposal was dropped when the PMM motor failed to demonstrate that it was ready to be installed in time.

Zumwalt will have Converteam's Advanced Induction Motors (AIM), rather than DRS Technologies' Permanent Magnet-Synchronous Motors (PMM).

> The exact choice of engine systems remains somewhat controversial at this point. The concept was originally for an integrated power system (IPS) based on in-hull permanent magnet synchronous motors (PMMs), with Advanced Induction Motors (AIM) as a possible backup solution. The design was shifted to the AIM system in February 2005 in order to meet scheduled milestones; PMM technical issues were subsequently fixed, but the program has moved on. The downside is that AIM technology has a heavier motor, requires more space, requires a "separate controller" to be developed to meet noise requirements, and produces one-third the amount of voltage. On the other hand, these very differences will force time and cost penalties from design and construction changes if the program wishes to "design AIM out" …[73]

The Integrated Power System (IPS) is, in some ways, similar to the old turbo-electric drive, the addition of PMMs and integration of all electrical power systems gives ten times the power available on current destroyers. It also reduces the ship's thermal and sound signature. The IPS has added to weight growth in the *Zumwalt*-class destroyer as noted by the GAO.[5]

Automation and fire protection

Automation reduces crew size on these ships: the *Zumwalt*-class destroyer's minimum complement is 130, less than half of needed by "similar warships",[48] Smaller crews reduce a major component of operating costs.[56] Ammunition, food, and other stores, are all mounted in containers able to be struck below to magazine/storage areas by an automated cargo handling system.[56]

Water spray or mist systems are proposed for deployment in the *Zumwalt*-class destroyer, but the electronic spaces remains problematic to the designers. Halon/Nitrogen dump systems are preferred but do not work when the space has been compromised by a hull breach. The GAO has noted this system as a potential problem yet to be addressed.[56][74]

Computer network

The Total Ship Computing Environment Infrastructure (TSCEI) is based on General Electric Fanuc Embedded Systems' PPC7A and PPC7D single-board computers[75] running LynuxWorks' LynxOS (Linux kernel)[76] RTOS.[77] These are contained in 16 shock, vibration and electromagnetic protected Electronic Modular Enclosures.[78] Zumwalt carries 16 pre-assembled IBM blade servers.[79]

1.2.4 Controversy

Lawmakers and others have questioned whether the *Zumwalt*-class costs too much and whether it provides the capabilities the U.S. military needs. In 2005 the Congressional Budget Office estimated the acquisition cost of a DD(X) at \$3.8–4.0bn in 2007 dollars, \$1.1bn more than the navy's estimate.[80]

The National Defense Authorization Act For Fiscal Year 2007 (Report of the Committee On Armed Services House of Representatives On H.R. 5122 Together With Additional And Dissenting Views) stated the following: "The committee understands there is no prospect of being able to design and build the two lead ships for the \$6.6 billion budgeted. The committee is concerned that the navy is attempting to insert too much capability into a single platform. As a result, the DD(X) is now expected to displace over 14,000 tons and by the navy's estimate, cost almost \$3.3 billion each. Originally, the navy proposed building 32 next generation destroyers, reduced that to 24, then to 7, and finally to 3, in order to make the program affordable. In such small numbers, the committee struggles to see how the original requirements for the next generation destroyer, for example providing naval surface fire support, can be met."

The following concerns have been raised about the design.

Ballistic missile/air defense capability

In January 2005, John Young, Assistant Secretary of the Navy for Research, Development and Acquisition, was so confident of the DD(X)'s improved air defense over the *Burke* class that between its new radar and ability to fire SM-1, SM-2, and SM-6, "I don't see as much urgency for [moving to] CG(X)" – a dedicated air defense cruiser.[81]

On 31 July 2008, Vice Admiral Barry McCullough (Deputy Chief of Naval Operations for Integration of Resources and Capabilities) and Allison Stiller (Deputy Assistant Secretary of the Navy for Ship Programs) stated that "the DDG 1000 cannot perform area air defense; specifically, it cannot successfully employ the Standard Missile-2 (SM-2), SM-

3 or SM-6 and is incapable of conducting Ballistic Missile Defense."[71] Dan Smith, president of Raytheon's Integrated Defense Systems division, has countered that the radar and combat system are essentially the same as other SM-2-capable ships, "I can't answer the question as to why the Navy is now asserting ... that *Zumwalt* is not equipped with an SM-2 capability".[23] The lack of anti-ballistic missile capability may represent a lack of compatibility with SM-2/SM-3. The *Arleigh Burke*-class ships have BMD systems with their Lockheed-Martin AEGIS tracking and targeting software,[82] unlike the DDG-1000's Raytheon TSCE-I targeting and tracking software,[75] which does not, as it is not yet complete, so while the DDG-1000, with its TSCE-I combat system, does have the SM-2/SM-3 missile system installed, it does not yet have the BMD/IAMD upgrade planned for the derived CG(X).[30] The Aegis system, on the other hand was used in the Aegis Ballistic Missile Defense System. Since the Aegis system has been the navy's chief combat system for the past 30 years when the navy started a BMD program, the combat system it was tested on was the Aegis combat system. So while the DDG-51 platform and the DDG-1000 platform are both SM-2/SM3 capable, as a legacy of the Aegis Ballistic Missile Defense System only the DDG-51 with the Aegis combat system is BMD capable, although the DDG-1000's TSCE-I combat system had both BMD and IAMD upgrades planned. And in view of recent intelligence that China is developing targetable anti-ship ballistic missiles based on the DF-21,[83][84] this could be a fatal flaw.

On 22 February 2009 James "Ace" Lyons, the former commander in chief of the U.S. Pacific Fleet, stated that the DDG-1000's technology was essential to a future "boost phase anti-ballistic missile intercept capability".[85]

In 2010, the Congressional Research Service reported that the DDG-1000 cannot currently be used for BMD because the BMD role was deferred to the DDG-1000 derived CG(X) program (The DDG's had the strike role, the CG had the BMD role, but they shared both the SM3 missile, and the TSCE-I), the proposed radar of the CG(X) was much larger (22')[86] and used much more energy and cooling capacity than the DDG-1000's.[30] Since then, the 22' (6.7 m) radar system has been canceled with the CG(X) and it has been determined that a 14' (4.2 m) radar could be used either on DDG-51 or on DDG-1000, though it would not have the performance the navy predicts would be needed "to address the most challenging threats".[86] Were the CG(X)'s BMD requirement adopted by the DDG-1000, the DDG-1000 would have to get the TSCE-I upgrade slated for the CG(X) to support that mission.[87]

The study that showed a cost benefit to building Flight III *Arleigh Burke*-class destroyer with enhanced radars instead of adding BMD to the *Zumwalt*-class destroyers assumed very limited changes from the Flight II to the Flight III

*Burke*s. However costs for the Flight III *Burke*s have increased rapidly "as the possible requirements and expectations continue to grow."[88] While the Flight III design and costs have been studied by the navy, there is very little reliable data available on what the cost would be to modify a DDG-1000–class ship to provide a BMD capability. However, if the Air Missile Defense Radar is adopted in common on both the Flight III *Burke*s and the *Zumwalt*s and if they were both upgraded to the same combat system then the only limitation of the *Zumwalt*s in this role would be their limited missile magazines.[89]

With the awarding of the development contract to the next generation Air and Missile Defense S-Band Radar to Raytheon, deliberation to put in place this radar on the *Zumwalt*-class destroyer is no longer being actively discussed.[90]

It is possible for the *Zumwalt*-class destroyers to get the more limited BMD hardware and software modifications that would allow them using their existing SPY-3 radar and Cooperative Engagement Capability to utilize the SM-3 missile and have a BMD capability similar to the BMD-capable *Ticonderoga*-class cruisers and *Burke*-class Flight IIa destroyers. Procurement of a BMD specific version of the *Zumwalt*-class destroyer is also being proposed.[30][91]

Zumwalt PLAS cells can launch the SM-2 Standard missile, but the ships have no requirement for ballistic missile defense. The tubes are long and wide enough to incorporate future interceptors, and although the ships' immediate role is littoral dominance and land attack, Raytheon contends that they could become BMD-capable with few modifications.[62]

Missile capacity

The original DD21 design, displacing around 16,000 tons, would have accommodated between 117 and 128 VLS cells.[92] However, the final DDG-1000 design was considerably smaller than that of the DD21, resulting in room for only 80 VLS cells.[93] Given the vessel's expected role, the *Zumwalt*-class destroyers will likely carry many more Tomahawk missiles than either the *Ticonderoga*- or *Arleigh Burke*-class ships.

Each VLS cell can be quad packed with RIM-162 Evolved Sea Sparrow Missiles (ESSM). This gives a maximum theoretical ESSM load out of 320 missiles. The ESSM is considered a point defense weapon not generally used for fleet area defense, although the ESSM has a range (27 NM) exceeding that of the earlier Naval Tartar anti-aircraft missile (17.5 NM RIM-24C).

Vice Admiral Barry McCullough On 31 July 2008 (deputy chief of naval operations for integration of resources and capabilities) and Allison Stiller (deputy assistant secretary of the navy for ship programs) stated that "the DDG 1000 cannot perform area air defense; specifically, it cannot successfully employ the Standard Missile-2 (SM-2), SM-3 or SM-6. It is not clear if the Standard Missile capability will be integrated into the *Zumwalt*-class destroyer or not.

The *Zumwalt*-class destroyer is not an Aegis system. It uses instead the class-unique Total Ship Computing Environment Infrastructure (TSCEI) integrated mission system. The peripheral vertical launch system (PVLS) VLS is capable of accommodating all Standard missile types [94] it has not been publicly stated if the TSCE will be modified to support the Standard missile or the ballistic missile defense mission.

Naval fire support role

Main article: United States Naval Gunfire Support debate

AGS being fired in September 2009 to test a new coating intended to extend barrel life at Dugway Proving Ground, Utah

> In summary, the committee is concerned that the navy has foregone the long range fire support capability of the battleship, has given little cause for optimism with regard to meeting near-term developmental objectives, and appears unrealistic in planning to support expeditionary warfare in the mid-term. The committee views the navy's strategy for providing naval surface fire support as 'high risk', and will continue to monitor progress accordingly.
> — Evaluation of the United States Navy's naval surface fire support program in the National Defense Authorization Act of 2007, [95]

A controversial point of the DD(X) destroyer(s) is their planned naval surface fire support (NSFS) role. The original DD21 and the Arsenal Ship had more serious NSFS capabilities, which would meet a Congress-mandated requirement related to the *Iowa*-class battleships. The requirement was eventually relaxed, the battleships stricken from the registry, and the navy left with small tonnage ships for NSFS or alternative methods such as air support. The official position of the U.S. Marine Corps and the U.S. Navy is that the *Zumwalt*-class destroyer(s) will be adequate as naval surface gunfire support ships, although there are dissenters.[96]

While smaller caliber guns (and missiles) have been used for centuries in naval fire support, very large guns have special capabilities beyond that of mid-range calibers. US battleships were re-activated three times after WWII specifically for naval fire support, and their 16 inch (406 mm) gunfire was used in every major engagement of the

U.S. from WWII through Operation Desert Storm in January/February 1991.[97] The *Zumwalt*-class will have two 6.1 inch (155 mm) guns with limited ammunition. The ships will fire a specially designed "guided" artillery shell some 63 nautical miles (117 km) inland.[95]

In March 2006, *Iowa* and *Wisconsin* were stricken from the Naval Vessel Register, having been kept on in part to fill a naval fire support role. However, Congress was "deeply concerned" over the loss of naval surface gunfire support they could provide and noted that "navy efforts to improve upon, much less replace, this capability have been highly problematic",[98] The U.S. House of Representatives asked that the battleships be kept in a state of readiness should they ever be needed again[99] and directed the navy to increase the number of *Arleigh Burke*-class destroyers that are currently being modernized.[99] The modernization includes extending the range of the 5-inch guns on the Flight 1 ships with Extended Range Guided Munitions (ERGMs) that would enable the ships to fire projectiles about forty nautical miles inland;[100][101] However the ERGM was canceled after it failed firing tests in February 2008.[102] The Navy is studying future options for naval fire support; Alliant Techsystems' ballistic trajectory extended range munition may be one possibility.[102] Adapting the 155 mm LRLAP to the 5"/54 Mk 45 gun is another option the navy is pursuing with BAE and Lockheed Martin as contractors.

Sea Jet, *out of the water and showing the unique hull design*

Sea Jet, *an Advanced Electric Ship Demonstrator*

Tumblehome design stability

The stability of the DDG-1000 hull design in heavy seas has been a matter of controversy. In April 2007, naval architect Ken Brower said, "As a ship pitches and heaves at sea, if you have tumblehome instead of flare, you have no righting energy to make the ship come back up. On the DDG 1000, with the waves coming at you from behind, when a ship pitches down, it can lose transverse stability as the stern comes out of the water – and basically roll over."[103] The decision to not use a tumblehome hull in the CG(X) cruiser, before the program was canceled, may suggest that there were concerns regarding *Zumwalt*'s seakeeping.[84] However, in a 1/4 scale test of the hull design, named *Sea Jet*, the tumblehome hull proved seaworthy.[104]

The Advanced Electric Ship Demonstrator (AESD), *Sea Jet*, funded by the Office of Naval Research (ONR), is a 133-foot (40-meter) vessel located at the Naval Surface Warfare Center Carderock Division, Acoustic Research Detachment in Bayview, Idaho. *Sea Jet* was operated on Lake Pend Oreille, where it was used for test and demonstration of various technologies. Among the first technologies tested was an underwater discharge water jet from

Rolls-Royce Naval Marine, Inc., called AWJ-21, a propulsion concept with the goals of providing increased propulsive efficiency, reduced acoustic signature, and improved maneuverability over previous destroyer-class combatants.

Secondary guns

In 2005, a Critical Design Review (CDR) of the DDG-1000 led to the selection of the Mk 110 57 mm cannon to defend the destroyer against swarming attacks by small fast-boats; the Mk 110 has a rate of fire of 220 rpm and a range of 9 nmi (10 mi; 17 km). From then to 2010, various analysis efforts were conducted to assess potential cost-saving alternatives. Following a 2012 assessment using the latest gun and munition effectiveness information, it was concluded that the Mk 46 30 mm Gun System was more effective than the Mk 110 with increased capability, reduced weight, and significant cost avoidance. The Mk 46 has a rate of fire of 200 rpm and a range of 2.17 nmi (2.50 mi; 4.02 km).[8]

Naval experts have questioned the decision to replace the close-in swarm defense guns of the *Zumwalt*-class destroyers with ones of decreased size and range. The 57 mm can engage targets at two to three miles, while the 30 mm can only start to engage at around one mile, inside the range of a rocket-propelled grenade fired from a small boat. However, the DDG-1000 program manager said that the 57 mm round's lethality was "significantly over-modeled" and "not as effective as modeled" in live test-firing, and "nowhere near meeting the requirements"; he admitted that the results were not what he expected to see. When the Naval Weapons Laboratory re-evaluated the Mk 46, it met or exceeded requirements and performed equal to or better than the 57 mm in multiple areas, even coming just ahead of the 76 mm naval cannon. A 30 mm gun mount also weighs less, around 2 tons compared to 12–14 tons for the 57 mm, but the navy is adamant that weight had nothing to do with the decision.[105]

1.2.5   Footnotes

[1] As of June 2014, the AGS the Zumwalt "can fire rocket-powered, computer-guided shells that can destroy targets 63 miles (101 km) away,... three times farther than ordinary destroyer guns can fire."

1.2.6   References

Citations

[1] "GAO-15-342SP DEFENSE ACQUISITIONS Assessments of Selected Weapon Programs" (PDF). US Government Accountability Office. March 2015. p. 73. Retrieved 15 July 2015.

[2] "Talking with the Chief Engineer Aboard DDG 1000". *CHIPS Magazine.* 23 September 2014.

[3] Destroyers – DDG fact file. U.S. Navy, 28 October 2009.

[4] Kasper, Joakim (20 September 2015). "About the Zumwalt Destroyer". *AeroWeb.* Retrieved 25 October 2015.

[5] *GAO-05-752R Progress of the DD(X) Destroyer Program.* U.S. Government Accountability Office. 14 June 2005.

[6] CRS RL32109 Navy DDG-51 and DDG-1000 Destroyer Programs: Background and Issues for Congress. CRS, 14 June 2010.

[7] "MK 57 Vertical Launch System". Raytheon

[8] Navy Swaps Out Anti-Swarm Boat Guns on DDG-1000s – News.USNI.org, 5 August 2014

[9] Section 1011 of the National Defense Authorization Act for Fiscal Year 1996 (Public Law 104-106; 110 Stat. 421)

[10] "New Zumwalt-Class Destroyer Is Not Your Father's Tin Can". *Los Angeles Times.* 5 July 2000.

[11] http://www.globalsecurity.org/military/library/news/2007/10/mil-071030-raytheon01.htm

[12] Sanchez, Lucia (January–March 2007). "Electromagnetic Railgun – A "Navy After Next" Game Changer". *CHIPS – the Department of the Navy Information Technology Magazine.* Retrieved 13 July 2013.

[13] "Boeing: Raygun dreadnoughts will rule the oceans by 2019". *The Register.* 17 April 2009. Retrieved 18 April 2009.

[14] NDAA 2007 pp. 69–70

[15] "Cutting-edge Navy warship being built in Maine". Fox News. 12 April 2012. Retrieved 12 April 2012.

[16] "Root Cause Analyses of Nunn-McCurdy Breaches, Volume 1" (PDF). RAND National Defense Research Institute. 2011. p. 19. Retrieved 2012-06-18.

[17] "Managing Affordability Industry Death Spiral UKUS". Strategy&

[18] Taylor, Andrew (26 September 2006). *House OKs $70B for Iraq, Afghanistan. The Washington Post.* Associated Press.

[19] *109th Congress :Department of Defense Appropriations Act, 2007.* (109–289) US Government Printing Office. 29 September 2006.

[20] "Navy: No Need to Add DDG 1000s After All". *Navy Times.* Gannett Government Media. 31 July 2008. Retrieved 25 January 2016.

[21] R. Jeffrey Smith and Ellen Nakashima. "Pentagon's Unwanted Projects in Earmarks". *Washington Post*, 8 March 2009. p. A01.

[22] Cavas, Christopher P (26 September 2008). "Roughead pushes for littoral combat ship". *Navy Times.*

[23] Cavas, Christopher P (16 September 2008). "Troubled DDG 1000 faces shipyard problems". *Navy Times.*

[24] Ewing, Philip (19 August 2008). *Lawmaker: Third DDG 1000 Far From Done Deal.* Defense News.

[25] Scully, Megan (24 September 2008). "Negotiators agree to buy more F-22s, Zumwalt destroyers". *Congress Daily.*

[26] Cavas, Christopher P., "New Destroyer Emerges in US Plans". *Defense News*, 2 February 2009, p. 1.

[27] Bennett, John T. and Kris Osborn. "Gates Reveals DoD Program Overhaul". *Defense News*, 6 April 2009.

[28] Drew, Christopher. "General Dynamics To Build New Destroyer", *New York Times*, 18 April 2009.

[29] John Pike. "CG(X) Next Generation Cruiser". Globalsecurity.org. Retrieved 2012-12-15.

[30] https://fas.org/sgp/crs/weapons/RL32109.pdf

[31] Galrahn (12 August 2008). "DDG-1000 and SM-2". Information Dissemination. Retrieved 2012-12-15.

[32] "What's Wrong with Weapons Acquisitions? – IEEE Spectrum". Spectrum.ieee.org. Retrieved 2012-12-15.

[33] "U.S. Navy Awards Northrop Grumman $90 Million Long-Lead Material Contract for DDG 1000". Northrop Grumman Corporation. 13 November 2007.

[34] "Navy Awards Contracts for Zumwalt Class Destroyers". Navy News Service. 14 February 2008.

[35] "BIW News February 2009" (PDF). General Dynamics Bath Iron Works. 1 March 2009.

[36] http://www.navsea.navy.mil/Newswire2011/17NOV11-01.aspx

[37] "America's Newest and Deadliest Destroyer Has Finally Set Sail". Gizmodo.com. Retrieved 2014-06-15.

[38] "GAO-08-804, Defense Acquisitions: Cost to Deliver Zumwalt-Class Destroyers Likely to Exceed Budget". Government Accountability Office. 31 July 2008.

[39] "First Zumwalt-Class Destroyer to Join U.S. Navy Fleet by late 2014". November 25, 2013.

[40] "Raytheon awarded $75 million for DDG 1000 Zumwalt class Destroyer program". December 18, 2013.

[41] "U.S. Navy Christened USS Zumwalt (DDG 1000), New Class of Destroyer". April 13, 2013.

[42] "Future Zumwalt class Destroyer USS Michael Monsoor (DDG 1001) Deckhouse Successfully Integrated". November 17, 2014.

[43] "General Dynamics Bath Iron Works Awarded $212 Million for DDG 1002 Deckhouse, Hangar and Launch-System Modules". August 6, 2013.

[44] DDG 1000 Preps for Heavy Weather Trials – DoD-Buzz.com, 14 January 2014

[45] Cindy Clayton (30 October 2008). "Navy to name newest destroyer after SEAL who died in Iraq". *The Virginian-Pilot*. Retrieved 22 April 2010.

[46] "Navy Names Zumwalt Class Destroyer USS Lyndon B. Johnson". Office of the Assistant Secretary of Defense (Public Affairs), United States Department of Defense. 16 April 2012. Retrieved 2012-04-16.

[47] "GAO Assessments of Major Weapon Programs." Government Accountability Office

[48] Patterson, Thom; Lendon, Brad (14 June 2014). "Navy's stealth destroyer designed for the video gamer generation". CNN. Retrieved 29 October 2014.

[49] "Zumwalt Class Destroyer Integrated Composite Deckhouse & Apertures (IDHA)". Raytheon Company. 22 March 2007.

[50] Cavas, Christopher P (12 September 2008). "Will DDG 1000 Produce Any Ships at All?". *DefenseNews*.

[51] Fabey, Michael (25 January 2013). "U.S. Navy Seeks Alternate Deckhouse For DDG-1002". *Aerospace Daily & Defense Report*.

[52] Schneider, David (31 July 2013). "The Electric Warship". *IEEE Spectrum*. IEEE. Retrieved 1 August 2013.

[53] "Navy Switches from Composite to Steel". Defense News, 2 August 2013.

[54] Forczyk. p. 18, 76

[55] Forczyk p. 32, 76

[56] "DDG-1000 Zumwalt / DD(X) Multi-Mission Surface Combatant". GlobalSecurity.org. 1 September 2008.

[57] "Wave Piercing Tumblehome Hull". Raytheon Company. 22 March 2007.

[58] Van Dam, L. Bruce (4 June 1999). *Does the Past Have a Place in the Future? The Utility of Battleships into the Twenty-First Century* (PDF). Fort Leavenworth, Kansas: US Army Command and General Staff College., citing a letter from Major Tracy Ralphs to Senator John Warner on 25 February 1999.

[59] "United States of America 16"/50 (40.6 cm) Mark 7". Tony DiGiulian, navweaps.com. 9 August 2008.

[60] "Advanced Gun System (AGS)". BAE Systems. 2008.

[61] "Zumwalt-Class Destroyer Critical Technologies". Raytheon.

[62] Ewing, Philip "SAS12: Approach of the Gray Elephant." *DoD Buzz*. 16 April 2012.

[63] "Zumwalt Class Destroyer Peripheral Vertical Launch System (PVLS) Advanced VLS". Raytheon Company. 22 March 2007.

[64] "Navy Switches from Composite to Steel | Defense News". defensenews.com. 2014-06-11. Retrieved 2014-06-15.

[65] "The US Navy's Dual Band Radars". Defenseindustrydaily.com. 11 August 2010. Retrieved 2011-12-27.

[66] http://www.navsea.navy.mil/nswc/dahlgren/Leading%20Edge/Sensors/03_Development.pdf

[67] http://www.dote.osd.mil/pub/reports/FY2011/pdf/navy/2011ssds.pdf

[68] O'Neil, William D. (August 2007). "The Cooperative Engagement Capability (CEC): Transforming Naval Anti-air Warfare" (PDF). Center for Technology and National Security Policy, National Defense University. Retrieved 29 October 2014.

[69] "Navy C4ISR and Unmanned Systems". *Sea Power 2016 Almanac*. Navy Leage of the U.S. January 2016. p. 91.

[70] Ronald O'Rourke (October 18, 2012). "Navy DDG-51 and DDG-1000 Destroyer Programs: Background and Issues for Congress" (PDF). Congressional Research Service. p. 23. Retrieved 2014-06-15.

[71] McCullough, Vice Adm. Barry; Stiller, Allison (31 July 2008). *Statement on Surface Combatant Requirements and Acquisition Strategy* (PDF). House Armed Services Committee.

[72] "Zumwalt Undersea Warfare Combat System Receives Official Navy Nomenclature". Raytheon. 9 December 2008.

[73] "Dead Aim, Or Dead End? The USA's DDG-1000 Zumwalt Class Program". *Defense Industry Daily*. 21 September 2008.

[74] "Zumwalt Class Destroyer Autonomic Fire Suppression System (AFSS)". Raytheon Company. 22 March 2007.

[75] "GE Fanuc Embedded Systems Selected By Raytheon For Zumwalt Class Destroyer Program". GE Fanuc Intelligent Platforms. 25 July 2007.

[76] http://www.techworm.net/2015/02/linux-used-build-us-navys-powerful-destroyer-yet.html

[77] "GE Fanuc Embedded Systems Selected By Raytheon For Zumwalt Class Destroyer Program". Lynuxworks. 25 July 2007.

[78] Gallagher, Sean (18 October 2013). "The Navy's newest warship is powered by Linux". Ars Technica. Retrieved 18 October 2013.

[79] http://www.techworm.net/2015/02/
linux-used-build-us-navys-powerful-destroyer-yet.html

[80] Gilmore, J. Michael (19 July 2005). *Statement on The Navy's DD(X) Destroyer Program before the Subcommittee on Projection Forces.* US House of Representatives.

[81] "John Young – Assistant Secretary of the US Navy For Research, Development And Acquisition". *Jane's Defence Weekly.* 12 January 2005. Archived from the original on 19 February 2009.

[82] "Aegis Ballistic Missile Defense". Mda.mil. 7 March 2011. Retrieved 2012-12-15.

[83] *Military Power of the People's Republic of China 2008* (PDF). Office of the Secretary of Defense. p. 2 (p12 of PDF).

[84] Cavas, Christopher P (4 August 2008). "Missile Threat Helped Drive DDG Cut". *DefenseNews.*

[85] LYONS: Naval shipbuilders sinking. *Washington Times,* 22 February 2009

[86] https://fas.org/sgp/crs/weapons/RL34179.pdf

[87] "CRS RL33745 Navy Aegis Ballistic Missile Defense (BMD) Program: Background and Issues for Congress 8 April 2010". Opencrs.com. Retrieved 2014-06-15.

[88] Fabey, Michael. "Potential DDG-51 Flight III Growth Alarms." *Aviation Week,* 10 June 2011.

[89] Cavas, Christopher P. "Axing DDG 1000 Radar May Save Cash, Enable BMD." *Defense News,* 4 June 2010.

[90] Christopher P. Cavas (October 10, 2013). "Raytheon Wins Key US Navy Radar Competition". defensenews.com. Retrieved 2014-06-15.

[91] "The US Navy's Dual Band Radars". *Defense Industry Daily.* 2013-10-01. Retrieved 2014-06-15.

[92] "DD-21 Zumwalt". globalsecurity.org. 27 April 2005.

[93] "DDG 1000 Flight I Design". Northrop Grumman Ship Systems. 2007. Archived from the original on 15 September 2007.

[94] "DDG1000_ASNE_Program_Overview_04.17.13" (PDF). Retrieved 2014-06-15.

[95] "National Defense Authorization Act of 2007" (PDF). p. 194. Retrieved 7 November 2008.

[96] Novak, Robert (6 December 2005). *Losing the Battleships.* CNN.

[97] AR 600-8-27 p. 26 paragraph 9–14

[98] NDAA 2007 p. 193

[99] NDAA 2007 p. 68

[100] NDAA 2007 pp. 67–68, 193

[101] "MK 45 5-inch / 54-caliber (lightweight) gun". Federation of American Scientists. 26 November 1999.

[102] Matthews, William (25 March 2008). "Navy ends ERGM funding". *Navy Times.*

[103] *Will DDG-1000 Destroyers Be Unstable?. Defense Industry Daily.* 12 April 2007., quoting Cavas, Christopher P (2 April 2007). "Is New U.S. Destroyer Unstable?". *DefenseNews.*

[104] John Pike. "Sea Jet Advanced Electric Ship Demonstrator (AESD)". Globalsecurity.org. Retrieved 2014-06-15.

[105] Experts Question US Navy's Decision To Swap Out DDG 1000's Secondary Gun – Defensenews.com, 12 October 2014

Bibliography

- Army Regulations 600-8-27 dated 2006

- Forczyk, Robert. *Russian Battleship vs Japanese Battleship, Yellow Sea 1904–05.* 2009 Osprey. ISBN 978-1-84603-330-8.

1.2.7 External links

- Raytheon's official DDG 1000 Program web page

- General DD(X) Destroyer page

- Description of the DD numbering system for ships in the U.S. Navy

- Overview of the DD(X) Destroyer program and its capabilities

- Zumwalt class Destroyer (Navy Recognition)

- Advanced Gun System set to be installed on the DD(X) destroyers

- DD(X) Destroyer program page on globalsecurity.org

- Concept of employment for naval surface fire support (near term capability)

- 1995 US General Accounting Office report on the US Navy's Naval Surface Fire Support program

- DoD press release: Navy Designates Next-Generation Zumwalt Destroyer

- Eaglen, Mackenzie (7 October 2008). "Changing Course on Navy Shipbuilding: Questions Congress Should Ask Before Funding". The Heritage Foundation.

- House letter recommending against a "winner take all" construction strategy for the DD(X) destroyer program

- Meet the Zumwalt: The Navy's stealth destroyer will go to sea next spring

- Canceling the DDG-1000 Destroyer Program Was a Mistake

Chapter 2

Features & Information

2.1 Advanced Gun System

The **Advanced Gun System** is a naval gun system developed and produced by BAE Systems Armaments Systems (formerly United Defense) for the *Zumwalt*-class destroyer of the United States Navy. The first magazine was delivered to the U.S. Navy on 25 May 2010.[1]

Originally designed for mounting as a vertical gun, this 155 millimetres (6.1 in) caliber gun has since been designed and produced for mounting within a more conventional turret arrangement. The AGS is designed to offer a weapon system capable of delivering precision munitions at a high rate of fire and at over-the-horizon ranges. As a vertical gun system it would only have been capable of firing guided munitions; the turret mounting will allow the use of unguided munitions as well.

The development of new ammunition for the AGS under the name Long Range Land Attack Projectile (LRLAP) is another major advance offered by the AGS program; it features separate projectile and propellant portions. The munitions are to be highly accurate, with a circular error probable (CEP) of 50 m (160 ft) or less. Lockheed Martin's flight test of the munition in July 2005 had a reported a flight distance of 59 nautical miles (109 km; 68 mi).

2.1.1 History

There has been research on extending the range of naval gunfire for many years. Gerald Bull and Naval Ordnance Station Indian Head tested an 11 in (280 mm) sub-caliber saboted long-range round[2] in a stretched 16"/45 caliber Mark 6 gun in 1967.[3] The Advanced Gun Weapon System Technology Program (AGWSTP) evaluated a similar projectile with longer range in the 1980s.[2] After the battleships were decommissioned in 1992, the AGWSTP became a 5-inch gun with an intended range of 180 km (110 mi), which then led to the Vertical Gun for Advanced Ships (VGAS). The original DD-21 was designed around this "vertical gun", but the project ran into serious technol-

ogy/cost problems and was radically scaled back to a more conventional 6.1 inch Advanced Gun System (AGS). One advantage of this move was that the gun was no longer restricted to guided munitions.

The naval gun system was developed and produced by BAE Systems Armaments Systems (formerly United Defense) for the *Zumwalt*-class destroyer of the United States Navy. The first magazine was delivered to the U.S. Navy on 25 May 2010.[4]

2.1.2 Description

The AGS uses the same 155 mm caliber as most American field artillery forces, although it is unable to fire the same ammunition. Instead, a new range of ammunition is under development for this weapon. The gun barrel is 62 calibers long, and is able to fire the entire magazine (300+ rounds) with an average rate of fire of ten rounds per minute using a water cooled barrel. The AGS is to be mounted in a turret specifically designed for the *Zumwalt* class destroyer with fully automated ammunition supply and operation. The turret itself is designed to be stealthy, allowing for the entire length of the barrel to be enclosed within the turret housing when not firing.

A primary advantage of the AGS over the existing Mark 45 5" gun which equips most major surface combatants of the US Navy is its increased capability for supporting ground forces and striking land targets. With a 10 round per minute capacity, it offers the ability to deliver firepower close to that of a battery of six 155 mm howitzers. This will increase the utility of vessels equipped with the weapon, especially in areas in which the US Navy exercises absolute sea supremacy.

2.1.3 Ammunition

The development of new ammunition for the AGS under the name Long Range Land Attack Projectile (LRLAP) is one of the major advances offered by the AGS program.

The munitions are to be highly accurate, with a circular error probable (CEP) of 50 m (160 ft) or less. Lockheed Martin conducted a flight test of the munition in July 2005, reporting a flight distance of 59 nautical miles (109 km; 68 mi).

The LRLAP ammunition features separate projectile and propellant portions. Total weight is 225 pounds (102 kg), including a bursting charge of 24 lb (11 kg). The maximum length of the combined munition is 88 in (220 cm), amounting to about 14 calibers.

2.1.4   See also

- 8"/55 caliber Mark 71 gun – US Navy's *Major Caliber Lightweight Gun (MCLWG)* program, designed & tested in 1975, program terminated in 1978.

2.1.5   Notes and references

[1] "BAE Systems Delivers First Piece of Production Hardware for U.S. Navy's Advanced Gun System". BAE Systems. 2010. Retrieved 2013-02-04.

[2] Van Dam, L. Bruce (1999-06-04). "Does the Past Have a Place in the Future? The Utility of Battleships into the Twenty-First Century" (PDF). Fort Leavenworth, Kansas: US Army Command and General Staff College., citing a letter from Major Tracy Ralphs to Senator John Warner on 1999-02-25

[3] "United States of America 16"/50 (40.6 cm) Mark 7". Tony DiGiulian, navweaps.com. 2008-08-09.

[4] "BAE Systems Delivers First Piece of Production Hardware for U.S. Navy's Advanced Gun System". BAE Systems. 2010. Retrieved 2013-02-04.

2.1.6   External links

- 155 mm/62 (6.1") Advanced Gun System (AGS)

Video links

- AGS employment in asymmetric warfare simulation scenario on YouTube

- AGS non-combatant evacuation simulation scenario on YouTube

2.2   Bath Iron Works

Warning: Page using Template:Infobox company with unknown parameter "origins" (this message is shown only in preview).

Bath Iron Works (**BIW**) is a major American shipyard

Bath Iron Works from NAS Brunswick photo gallery

located on the Kennebec River in Bath, Maine. Since its founding in 1884 (as Bath Iron Works, Limited), BIW has built private, commercial and military vessels, most of which have been ordered by the United States Navy. The shipyard has built and sometimes designed battleships, frigates, cruisers and destroyers, including the *Arleigh Burke* class, which are currently among the world's most advanced surface warships.

Since 1995, Bath Iron Works has been a subsidiary of General Dynamics, the fifth-largest defense contractor in the world (as of 2008). During World War II, ships built at BIW were considered by sailors and Navy officials to be of superior toughness, giving rise to the phrase *"Bath-built is best-built."* [1]

2.2.1   History

Bath Iron Works was incorporated in 1884 by General Thomas W. Hyde, a native of Bath who served in the American Civil War. After the war, Hyde bought a local shop that helped make windlasses and other iron hardware for the wooden ships built in Bath's many shipyards. He expanded the business by improving its practices, entering new markets, and acquiring other local businesses.

By 1882, Hyde Windlass was eyeing the new and growing business of iron shipbuilding; two years later, it incorporated as Bath Iron Works. On February 28, 1890, BIW won its first contract for complete vessels, two iron gunboats for the U.S. Navy. The *Machias*, one of these 190-foot (58 m) gunboats, was the first ship launched by the company. (Historian Snow (see "Further Reading") says the gunboat was commanded during World War I by Chester Nimitz, an assertion that is not supported by Nimitz's biographers.)

In 1892, the yard won its first commercial contract for a

steel vessel, the 2,500-ton steel passenger steamer *City of Lowell*. In the 1890s, the company built several yachts for wealthy sailors.

In 1899, General Hyde, suffering from the Bright's Disease that would kill him later that year, resigned from management of the shipyard, leaving his sons Edward and John in charge. That year the shipyard began construction of the *Georgia*, the only battleship to be built in Bath. The ship dominated the yard for five years until its launching in 1904, and was at times the only ship under construction. The yard faced numerous challenges because of the weight of armor and weapons. In sea trials, the *Georgia* averaged 19.26 knots (35.67 km/h) for four hours, making her the fastest ship in her class and the fastest battleship in the Navy.

The company continued to rely on Navy contracts, which provided 86% of the value of new contracts between 1905 and 1917. The yard also produced fishing trawlers, freighters, and yachts throughout the first half of the century.

At peak production during World War II (1943–1944), the shipyard launched a destroyer every 17 days. Bath Iron Works ranked 50th among United States corporations in the value of World War II military production contracts.[2]

In 1981, Falcon Transport ordered two tankers, the last commercial vessels built by BIW.

MV Mighty Servant 2 *carrying mine-damaged* Roberts *on 31 July 1988*

In 1988, the USS *Samuel B. Roberts* (FFG-58), commissioned two years earlier at Bath, survived a mine explosion that tore a hole in its engine room and flooded two compartments. Over the next two years, BIW repaired the *Roberts* in unique fashion. The guided missile frigate was towed to the company's dry dock in Portland, Maine, and put up on blocks, where its damaged engine room was cut out of the ship. Meanwhile, workers in Bath built a 315-ton replacement. When it was ready, the module was floated south to Portland, placed on the dry dock, slid into place under the *Roberts*, jacked up, and welded into place.[3] By surviving a hit that Naval Sea Systems Command engineers thought should have sunk her, the *Roberts* validated the penny-pinching design of the *Oliver Hazard Perry* class, the U.S. Navy's largest post-WWII class until the *Burkes* ; and validated the Navy's against-the-odds decision to have picked BIW to design it.

In 2001, BIW wrapped up a four-year effort to build an enormous concrete platform, the Land Level Transfer Facility, for final assembly of its ships. Instead of being built on a sloping way so that they could slide into the Kennebec at launch, hulls were henceforth moved by rail from the platform horizontally onto a moveable dry dock. This greatly reduced the work involved in building and launching the ships.[4] The 750-foot, 28,000-ton dry dock was built by China's Jiangdu Yuchai Shipbuilding Company for $27 million.[5]

The Centennial Shipbuilders Workers Monument in Bath, Maine is by American artist Guillermo Esparza and is part of the Smithsonian American Art Museum collection.

2.2.2 Notable ships built

USS Chester *(CL-1) was the first United States cruiser of the numbering series used through the first half of the 20th century.*

- Yachts

  - Ranger, successful America's Cup defender

The last of the "four-stack" destroyers, USS Pruitt *(DD-347) being launched from Bath Iron Works in 1920.*

- Aras II, Presidential Yacht known as USS Williamsburg

- Lightvessels
    - Diamond Shoal Lightship No. 71 (LV-71)
    - Nantucket Lightship 66
    - Nantucket Lightship 106

- Naval ram
    - USS *Katahdin*

- Monitor
    - USS *Nevada* (BM-8)[6]

- *Denver* class protected cruiser
    - USS *Cleveland* (C-19) World War I

- *Virginia*-class battleship
    - USS *Georgia* (BB-15), launched in 1904

- *Chester*-class cruiser
    - USS *Chester* (CL-1) World War I

- *Smith*-class destroyers
    - USS *Flusser* (DD-20) World War I
    - USS *Reid* (DD-21) World War I

- *Paulding*-class destroyers
    - USS *Paulding* (DD-22) World War I - Rum Patrol
    - USS *Drayton* (DD-23) World War I
    - USS *Trippe* (DD-33) World War I - Rum Patrol

- USS *Jouett* (DD-41) World War I - Rum Patrol
- USS *Jenkins* (DD-42) World War I

- *Cassin*-class destroyers
    - USS *Cassin* (DD-43) World War I - Rum Patrol
    - USS *Cummings* (DD-44) World War I - Rum Patrol

- *O'Brien*-class destroyer
    - USS *McDougal* (DD-54) World War I - Rum Patrol

- *Tucker*-class destroyer
    - USS *Wadsworth* (DD-60) World War I

- *Sampson*-class destroyers
    - USS *Davis* (DD-65) World War I - Rum Patrol
    - USS *Allen* (DD-66)[7] World War I - Attack on Pearl Harbor

- *Caldwell*-class destroyer
    - USS *Manley* (DD-74)[8] World War I - Guadalcanal Campaign - Operation Flintlock - Battle of Saipan - Philippines campaign (1944-45)

Two of the seven Bath Iron Works destroyers transferred to the Royal Navy in the Destroyers for Bases Agreement. The outboard ship made the St. Nazaire Raid.

- *Wickes*-class destroyers
    - USS *Wickes* (DD-75)[9] World War I - Destroyers for Bases Agreement
    - USS *Philip* (DD-76)[9] World War I - Destroyers for Bases Agreement

- USS *Woolsey* (DD-77)[9] World War I
- USS *Evans* (DD-78)[9] Destroyers for Bases Agreement
- USS *Buchanan* (DD-131)[9] Destroyers for Bases Agreement - St. Nazaire Raid
- USS *Aaron Ward* (DD-132)[9] Destroyers for Bases Agreement
- USS *Hale* (DD-133)[9] Destroyers for Bases Agreement
- USS *Crowninshield* (DD-134)[9] Destroyers for Bases Agreement

- *Clemson*-class destroyers

  - USS *Preble* (DD-345)[10] Attack on Pearl Harbor - Guadalcanal Campaign
  - USS *Sicard* (DD-346)[10] Attack on Pearl Harbor - Battle of Empress Augusta Bay
  - USS *Pruitt* (DD-347)[10] Attack on Pearl Harbor

USCGC Icarus (WPC-110) delivers prisoners from U-352 to Charleston Navy Yard on 10 May 1942.

- *Thetis*-class patrol boat

  - USCGC *Aurora* (WPC-103)[11]
  - USCGC *Calypso* (WPC-104)[11]
  - USCGC *Daphne* (WPC-106)[12]
  - USCGC *Hermes* (WPC-109)[12]
  - USCGC *Icarus* (WPC-110)[12] sank *U-352*
  - USCGC *Perseus* (WPC-114)[12]
  - USCGC *Thetis* (WPC-115)[12] sank *U-157*

- *Farragut*-class destroyers (1934)

- USS *Dewey* (DD-349)[13] Attack on Pearl Harbor - Battle of the Coral Sea[14] - Battle of Midway - Guadalcanal Campaign - Battle of the Eastern Solomons - Battle of the Philippine Sea[15]

- The J-class yacht *Ranger*, 1936

- *Mahan*-class destroyers

  - USS *Drayton* (DD-366)[16] Battle of Tassafaronga[17] Philippines campaign (1944-45)
  - USS *Lamson* (DD-367)[16] Battle of Tassafaronga[17] - Philippines campaign (1944-45) - sunk in test *Able* of Operation Crossroads

- *Somers*-class destroyers

  - USS *Sampson* (DD-394)[16]
  - USS *Davis* (DD-395)[16]
  - USS *Jouett* (DD-396)[16] Invasion of Normandy

- *Sims*-class destroyers

  - USS *Sims* (DD-409)[18] Battle of the Coral Sea[19]
  - USS *Hughes* (DD-410)[18] Battle of Midway[20] - Battle of the Santa Cruz Islands[21] - Naval Battle of Guadalcanal[22] - Philippines campaign (1944-45)

- *Gleaves*-class destroyers

  - USS *Gleaves* (DD-423)[18] invasions of Sicily, Italy and Southern France
  - USS *Niblack* (DD-424)[18] invasions of Sicily, Italy and Southern France
  - USS *Livermore* (DD-429)[23] invasions of North Africa and Southern France
  - USS *Eberle* (DD-430)[23] invasions of North Africa and Southern France
  - USS *Woolsey* (DD-437)[23] invasions of North Africa, Sicily and Italy
  - USS *Ludlow* (DD-438)[23] invasions of North Africa, Sicily, Italy and Southern France
  - USS *Emmons* (DD-457)[24] invasions of North Africa, Normandy, Southern France and Okinawa
  - USS *Macomb* (DD-458)[24] invasions of North Africa, Southern France and Okinawa

- *Fletcher*-class destroyers

Nicholas holds the United States Navy record for battle stars with 16 from World War II, 5 from the Korean War and 9 from the Vietnam War

- USS *Nicholas* (DD-449)[25] Guadalcanal campaign - Philippines campaign (1944-45) - Korean War - Vietnam War

- USS *O'Bannon* (DD-450)[25] Naval Battle of Guadalcanal[26] Guadalcanal campaign - Naval Battle of Vella Lavella[27] - Philippines campaign (1944-45) - Korean War - Vietnam War

- USS *Chevalier* (DD-451)[25] Guadalcanal campaign - Naval Battle of Vella Lavella[27]

- USS *Strong* (DD-467)[25] Guadalcanal campaign

- USS *Taylor* (DD-468)[25] Guadalcanal campaign - Philippines campaign (1944-45) - Korean War - Vietnam War

- USS *De Haven* (DD-469)[25] Guadalcanal campaign

- USS *Conway* (DD-507)[28] Guadalcanal campaign - Philippines campaign (1944-45) - Korean War

- USS *Cony* (DD-508)[28] Guadalcanal campaign - Philippines campaign (1944-45) - Battle of Surigao Strait - Korean War

- USS *Converse* (DD-509)[28] Guadalcanal campaign - Battle of Empress Augusta Bay[29] Battle of Cape St. George[30] - Battle of the Philippine Sea[15] - Philippines campaign (1944-45)

- USS *Eaton* (DD-510)[28] Guadalcanal campaign - Philippines campaign (1944-45)

- USS *Foote* (DD-511)[28] Guadalcanal campaign - Battle of Empress Augusta Bay[29] - Philippines campaign (1944-45) - Battle of Okinawa

- USS *Spence* (DD-512)[28] Guadalcanal campaign - Battle of Empress Augusta Bay[29] - Battle of Cape St. George[30] - Battle of the Philippine Sea[15] - Philippines campaign (1944-45)

- USS *Terry* (DD-513)[28] Guadalcanal campaign - Battle of the Philippine Sea[15] - Battle of Iwo Jima

- USS *Thatcher* (DD-514)[28] Guadalcanal campaign - Battle of Empress Augusta Bay[29] - Battle of the Philippine Sea[15] - Philippines campaign (1944-45) - Battle of Okinawa

- USS *Anthony* (DD-515)[28] Guadalcanal campaign - Battle of the Philippine Sea[15] - Battle of Okinawa

- USS *Wadsworth* (DD-516)[28] Guadalcanal campaign - Battle of the Philippine Sea[15] - Philippines campaign (1944-45) - Battle of Okinawa

- USS *Walker* (DD-517)[28] Philippines campaign (1944-45) - Battle of Okinawa - Korean War - Vietnam War

- USS *Abbot* (DD-629)[31] Philippines campaign (1944-45)

- USS *Braine* (DD-630)[31] Battle of the Philippine Sea[15] - Philippines campaign (1944-45) - Battle of Okinawa

- USS *Erben* (DD-631)[31] Philippines campaign (1944-45) - Battle of Okinawa - Korean War

- USS *Hale* (DD-642)[31] Philippines campaign (1944-45) - Battle of Okinawa

- USS *Sigourney* (DD-643)[31] Guadalcanal campaign - Philippines campaign (1944-45) - Battle of Surigao Strait

- USS *Stembel* (DD-644)[31] Philippines campaign (1944-45) - Battle of Okinawa - Korean War

- USS *Caperton* (DD-650)[31] Battle of the Philippine Sea[15] - Philippines campaign (1944-45)

- USS *Cogswell* (DD-651)[31] Battle of the Philippine Sea[15] - Philippines campaign (1944-45) - Vietnam War

- USS *Ingersoll* (DD-652)[31] Philippines campaign (1944-45)[15] - Vietnam War

- USS *Knapp* (DD-653)[31] Battle of the Philippine Sea[15] - Philippines campaign (1944-45)

- USS *Remey* (DD-688)[32] Battle of Saipan - Philippines campaign (1944-45) - Battle of Surigao Strait - Battle of Okinawa

- USS *Wadleigh* (DD-689)[32] Battle of Saipan

- USS *Norman Scott* (DD-690)[32] Battle of Saipan

- USS *Mertz* (DD-691)[32] Philippines campaign (1944-45)

- *Allen M. Sumner*-class destroyers

- USS *Barton* (DD-722)[33] Invasion of Normandy - Philippines campaign (1944-45) - Korean War

- USS *Walke* (DD-723)[33] Invasion of Normandy - Philippines campaign (1944-45) - Battle of Okinawa - Korean War - Vietnam War

- USS *Laffey* (DD-724)[33] Invasion of Normandy - Philippines campaign (1944-45) - Battle of Okinawa - Korean War - preserved National Historic Landmark in Charleston, South Carolina

- USS *O'Brien* (DD-725)[33] Invasion of Normandy - Philippines campaign (1944-45) - Korean War - Vietnam War

- USS *Meredith* (DD-726)[33] Invasion of Normandy

- USS *De Haven* (DD-727)[33] Philippines campaign (1944-45) - Battle of Okinawa - Korean War

- USS *Mansfield* (DD-728)[33] Philippines campaign (1944-45) - Korean War - Vietnam War

- USS *Lyman K. Swenson* (DD-729)[33] Philippines campaign (1944-45) - Battle of Okinawa - Korean War - Vietnam War

- USS *Collett* (DD-730)[33] Philippines campaign (1944-45) - Korean War

- USS *Maddox* (DD-731)[33] Battle of Okinawa - Korean War - Gulf of Tonkin Incident - Vietnam War

- USS *Hyman* (DD-732)[33] Battle of Okinawa - Korean War

- USS *Mannert L. Abele* (DD-733)[33] Battle of Okinawa

- USS *Purdy* (DD-734)[33] Battle of Okinawa - Korean War

- USS *Robert H. Smith* (DM-23)[10] Battle of Okinawa

- USS *Thomas E. Fraser* (DM-24)[10] Battle of Okinawa

- USS *Shannon* (DM-25)[10] Battle of Okinawa

- USS *Harry F. Bauer* (DM-26)[10] Battle of Okinawa

- USS *Adams* (DM-27)[10] Battle of Okinawa

- USS *Tolman* (DM-28)[10] Battle of Okinawa

- USS *Drexler* (DD-741)[33] Battle of Okinawa

- *Gearing*-class destroyers

  - USS *Frank Knox* (DD-742)[34] World War II - Korean War - Vietnam War

Agerholm launched an ASROC anti-submarine rocket armed with a nuclear depth bomb during the Swordfish test of 1962

- USS *Southerland* (DD-743)[34] World War II - Korean War - Vietnam War

- USS *Chevalier* (DD-805)[35] Korean War

- USS *Higbee* (DD-806)[35] World War II - Korean War - Vietnam War - Battle of Dong Hoi

- USS *Benner* (DD-807)[35] World War II - Vietnam War

- USS *Dennis J. Buckley* (DD-808)[35] Vietnam War

- USS *Agerholm* (DD-826)[35] Korean War - Vietnam War

- USS *Robert A. Owens* (DD-827)[35]

- USS *Timmerman* (DD-828)[35] (Experimental ship completed with aluminum superstructure and high-horsepower engines)

- USS *Myles C. Fox* (DD-829)[35] Vietnam War

- USS *Everett F. Larson* (DD-830)[35] Vietnam War

- USS *Goodrich* (DD-831)[35]

- USS *Hanson* (DD-832)[35] Korean War - Vietnam War

- USS *Herbert J. Thomas* (DD-833)[35] Korean War - Vietnam War

- USS *Turner* (DD-834)[35]

- USS *Charles P. Cecil* (DD-835)[35] Vietnam War

- USS *George K. MacKenzie* (DD-836)[35] Korean War - Vietnam War

- USS *Sarsfield* (DD-837)[35] Vietnam War

- USS *Ernest G. Small* (DD-838)[35] Korean War

- USS *Power* (DD-839)[35] Vietnam War

- USS *Glennon* (DD-840)[35]

- USS *Noa* (DD-841)[35] Recovered astronaut John Glenn in Friendship 7 on 20 February 1962

- USS *Fiske* (DD-842)[35] Korean War - Vietnam War

- USS *Warrington* (DD-843)[35]

- USS *Perry* (DD-844)[35] Vietnam War

- USS *Bausell* (DD-845)[35] Korean War - Vietnam War

- USS *Ozbourn* (DD-846)[35] Korean War - Vietnam War

- USS *Robert L. Wilson* (DD-847)[35] Vietnam War

- USS *Witek* (DD-848)[36] (no overseas deployments - used exclusively for ASW research)

- USS *Richard E. Kraus* (DD-849)[36] Vietnam War

- *Dealey*-class destroyer escorts

  - USS *Dealey* (DE-1006)[37]

  - USS *Cromwell* (DE-1014)[37]

  - USS *Hammerberg* (DE-1015)[37]

The second Cold War destroyer built by Bath Iron Works was named for the grandfather of Republican 2008 presidential candidate John S. McCain III.

- *Mitscher*-class destroyers

  - USS *Mitscher* (DL-2)[38]

  - USS *John S. McCain* (DL-3)[38] Vietnam War

- *Forrest Sherman*-class destroyers

  - USS *Forrest Sherman* (DD-931)[39]

- USS *John Paul Jones* (DD-932)[39]

- USS *Barry* (DD-933)[39] Vietnam War

- USS *Manley* (DD-940)[39] Vietnam War

- USS *Dupont* (DD-941)[39]

- USS *Bigelow* (DD-942)[39] Vietnam War

- USS *Hull* (DD-945)[39] Vietnam War

- USS *Edson* (DD-946)[39] Vietnam War

- USS *Somers* (DD-947)[39] Vietnam War

- *Charles F. Adams*-class destroyers

  - USS *Charles F. Adams* (DDG-2)[40]

  - USS *John King* (DDG-3)[40]

  - USS *Sampson* (DDG-10)[40]

  - USS *Sellers* (DDG-11)[40]

- *Farragut*-class destroyers (1958)

  - USS *Dewey* (DDG-45)[41]

  - USS *Preble* (DDG-46)[41] Vietnam War

- *Leahy*-class cruisers

  - USS *Leahy* (CG-16)[42]

  - USS *Harry E. Yarnell* (CG-17)[42]

  - USS *Worden* (CG-18)[42] Vietnam War

- *Belknap*-class cruisers

  - USS *Belknap* (CG-26)[43]

  - USS *Josephus Daniels* (CG-27)[43]

  - USS *Wainwright* (CG-28)[43] Vietnam War

  - USS *William H. Standley* (CG-32)[43] Vietnam War

  - USS *Biddle* (CG-34)[43] Vietnam War

- *Garcia*-class frigate

  - USS *Glover* (FF-1098)[44]

- *Brooke*-class frigates

  - USS *Talbot* (FFG-4)[45]

  - USS *Richard L. Page* (FFG-5)[45]

  - USS *Julius A. Furer* (FFG-6)[45]

- *Oliver Hazard Perry*-class frigates

  - USS *Oliver Hazard Perry* (FFG-7)[46]

  - USS *McInerney* (FFG-8)[46]

  - USS *Clark* (FFG-11)[46]

  - USS *Samuel Eliot Morison* (FFG-13)[46]

- USS *Estocin* (FFG-15)[46]
- USS *Clifton Sprague* (FFG-16)[46]
- USS *Flatley* (FFG-21)[46]
- USS *Jack Williams* (FFG-24)[46]
- USS *Gallery* (FFG-26)[46]
- USS *Stephen W. Groves* (FFG-29)[46]
- USS *John L. Hall* (FFG-32)[46]
- USS *Aubrey Fitch* (FFG-34)[46]
- USS *Underwood* (FFG-36)[46]
- USS *Doyle* (FFG-39)[46]
- USS *Klakring* (FFG-42)[46]
- USS *Dewert* (FFG-45)[46]
- USS *Nicholas* (FFG-47)[46]
- USS *Robert G. Bradley* (FFG-49)[46]
- USS *Taylor* (FFG-50)
- USS *Hawes* (FFG-53)
- USS *Elrod* (FFG-55)
- USS *Simpson* (FFG-56), launched August 31, 1984. One of four U.S. Navy ships in commission to have sunk an enemy vessel with shipboard weaponry, the others being the USS *Constitution*, USS *Porter* (DDG-78), and USS *Carter Hall* (LSD-50),
- USS *Samuel B. Roberts* (FFG-58), launched in 1984 and repaired after being punctured by a mine in 1988
- USS *Kauffman* (FFG-59)

- *Ticonderoga*-class cruisers

  - USS *Thomas S. Gates* (CG-51)
  - USS *Philippine Sea* (CG-58)
  - USS *Normandy* (CG-60)
  - USS *Monterey* (CG-61)
  - USS *Cowpens* (CG-63)
  - USS *Gettysburg* (CG-64)
  - USS *Shiloh* (CG-67)
  - USS *Lake Erie* (CG-70), 21 Feb 2008 shot down the errant USA 193 satellite with a modified SM3 missile.

- *Arleigh Burke*-class destroyers

  - USS *Arleigh Burke* (DDG-51), commissioned July 4, 1991.
  - USS *John Paul Jones* (DDG-53)
  - USS *Curtis Wilbur* (DDG-54)

  - USS *John S. McCain* (DDG-56)
  - USS *Laboon* (DDG-58)
  - USS *Paul Hamilton* (DDG-60)
  - USS *Fitzgerald* (DDG-62)
  - USS *Carney* (DDG-64)
  - USS *Gonzalez* (DDG-66)
  - USS *The Sullivans* (DDG-68)
  - USS *Hopper* (DDG-70)
  - USS *Mahan* (DDG-72)
  - USS *Decatur* (DDG-73)
  - USS *Donald Cook* (DDG-75)
  - USS *Higgins* (DDG-76)
  - USS *O'Kane* (DDG-77)
  - USS *Oscar Austin* (DDG-79)
  - USS *Winston S. Churchill* (DDG-81)
  - USS *Howard* (DDG-83)
  - USS *McCampbell* (DDG-85)
  - USS *Mason* (DDG-87)
  - USS *Chafee* (DDG-90)
  - USS *Momsen* (DDG-92)
  - USS *Nitze* (DDG-94)
  - USS *Bainbridge* (DDG-96), launched in 2005
  - USS *Farragut* (DDG-99)
  - USS *Gridley* (DDG-101), launched in 2006
  - USS *Sampson* (DDG-102)
  - USS *Sterett* (DDG-104)
  - USS *Stockdale* (DDG-106)
  - USS *Wayne E. Meyer* (DDG-108)
  - USS *Jason Dunham* (DDG 109)
  - USS *Michael Murphy* (DDG-112)

- *Zumwalt*-class destroyers

  - USS *Zumwalt* (DDG-1000)

2.2.3 References

[1] See Peniston, Sanders, Snow.

[2] Peck, Merton J. & Scherer, Frederic M. *The Weapons Acquisition Process: An Economic Analysis* (1962) Harvard Business School p.619

[3] No Higher Honor: FFG 58 Repair

[4] GDBIW.com

[5] "Bath Iron Works picks Chinese firm". United Press International. 1998-09-14. Retrieved 2008-10-18.

[6] "Nevada". *Dictionary of American Naval Fighting Ships.* United States Navy. Retrieved 12 December 2013.

[7] Silverstone, Paul H. *U.S. Warships of World War II* Doubleday & Company (1968) p.103

[8] Silverstone, Paul H. *U.S. Warships of World War II* Doubleday & Company (1968) p.276

[9] Fahey, James C. *The Ships and Aircraft of the United States Fleet* Ships and Aircraft (1939) p.17

[10] Silverstone, Paul H. *U.S. Warships of World War II* Doubleday & Company (1968) p.212

[11] Silverstone, Paul H. *U.S. Warships of World War II* Doubleday & Company (1968) p.380

[12] Silverstone, Paul H. *U.S. Warships of World War II* Doubleday & Company (1968) p.383

[13] Silverstone, Paul H. *U.S. Warships of World War II* Doubleday & Company (1968) p.114

[14] Oftsie, R.A., RADM USN *The Campaigns of the Pacific War* United States Government Printing Office (1946) p.55

[15] Tillman, Barrett *Clash of the Carriers* (2005) ISBN 0-451-21956-2 pp.301-306

[16] Silverstone, Paul H. *U.S. Warships of World War II* Doubleday & Company (1968) p.118

[17] Oftsie, R.A., RADM USN *The Campaigns of the Pacific War* United States Government Printing Office (1946) p.140

[18] Silverstone, Paul H. *U.S. Warships of World War II* Doubleday & Company (1968) p.126

[19] Oftsie, R.A., RADM USN *The Campaigns of the Pacific War* United States Government Printing Office (1946) p.54

[20] Oftsie, R.A., RADM USN *The Campaigns of the Pacific War* United States Government Printing Office (1946) p.74

[21] Oftsie, R.A., RADM USN *The Campaigns of the Pacific War* United States Government Printing Office (1946) p.122

[22] Oftsie, R.A., RADM USN *The Campaigns of the Pacific War* United States Government Printing Office (1946) p.128

[23] Silverstone, Paul H. *U.S. Warships of World War II* Doubleday & Company (1968) p.129

[24] Silverstone, Paul H. *U.S. Warships of World War II* Doubleday & Company (1968) p.132

[25] Silverstone, Paul H. *U.S. Warships of World War II* Doubleday & Company (1968) p.135

[26] Oftsie, R.A., RADM USN *The Campaigns of the Pacific War* United States Government Printing Office (1946) p.127

[27] Oftsie, R.A., RADM USN *The Campaigns of the Pacific War* United States Government Printing Office (1946) p.148

[28] Silverstone, Paul H. *U.S. Warships of World War II* Doubleday & Company (1968) p.138

[29] Oftsie, R.A., RADM USN *The Campaigns of the Pacific War* United States Government Printing Office (1946) p.153

[30] Oftsie, R.A., RADM USN *The Campaigns of the Pacific War* United States Government Printing Office (1946) p.159

[31] Silverstone, Paul H. *U.S. Warships of World War II* Doubleday & Company (1968) p.141

[32] Silverstone, Paul H. *U.S. Warships of World War II* Doubleday & Company (1968) p.143

[33] Silverstone, Paul H. *U.S. Warships of World War II* Doubleday & Company (1968) pp.146-7

[34] Silverstone, Paul H. *U.S. Warships of World War II* Doubleday & Company (1968) p.148

[35] Silverstone, Paul H. *U.S. Warships of World War II* Doubleday & Company (1968) p.150

[36] Silverstone, Paul H. *U.S. Warships of World War II* Doubleday & Company (1968) p.152

[37] Blackman, Raymond V. B. *Jane's Fighting Ships* (1970/71) p.458

[38] Blackman, Raymond V. B. *Jane's Fighting Ships* (1970/71) p.435

[39] Blackman, Raymond V. B. *Jane's Fighting Ships* (1970/71) p.439

[40] Blackman, Raymond V. B. *Jane's Fighting Ships* (1970/71) p.437

[41] Blackman, Raymond V. B. *Jane's Fighting Ships* (1970/71) p.432

[42] Blackman, Raymond V. B. *Jane's Fighting Ships* (1970/71) p.431

[43] Blackman, Raymond V. B. *Jane's Fighting Ships* (1970/71) p.429

[44] Blackman, Raymond V. B. *Jane's Fighting Ships* (1970/71) p.456

[45] Blackman, Raymond V. B. *Jane's Fighting Ships* (1970/71) p.452

[46] Clement, Janet Ann, LT USNR "The FFG-7 Program: A Shipbuilding Status Report" *United States Naval Institute Proceedings* (June 1981) p.109

2.2.4 Further reading

- Eskew, Garnett Laidlaw (1958). *Cradle of Ships*. New York: Putnam. ASIN B0007E5VY4. (First general history of BIW.)

- Peniston, Bradley (2006). *No Higher Honor: Saving the USS Samuel B. Roberts in the Persian Gulf*. Annapolis: Naval Institute Press. ISBN 1-59114-661-5. (Describes the construction of a *Perry*-class guided missile frigate, the training of its precommissioning crew at BIW, and the complex repair job that returned it to duty.)

- Sanders, Michael S. (1999). *The Yard: Building a Destroyer at the Bath Iron Works*. New York: Harper-Collins. ISBN 0-06-019246-1. (Describes the construction of USS *Donald Cook* (DDG-75) at BIW.)

- Snow, Ralph L. (1987). *Bath Iron Works: The First Hundred Years*. Bath, Maine: Maine Maritime Museum. ISBN 0-9619449-0-0. (The definitive work on BIW from 1884-1987.)

- Toppan, Andrew (2002). *Bath Iron Works (Images of America: Maine)*. South Carolina: Arcadia Publishing. ISBN 0-7385-1059-9. (Historic and contemporary photos of BIW.)

2.2.5 External links

- Bath Iron Works website

- USS *Samuel B. Roberts* (FFG-58) under repair at BIW's Portland dry dock

Coordinates: 43°54′16″N 69°48′53″W / 43.904494°N 69.814746°W

2.3 Elmo Zumwalt

Elmo Russell "Bud" Zumwalt, Jr. (November 29, 1920 – January 2, 2000) was an American naval officer and the youngest man to serve as Chief of Naval Operations. As an admiral and later the 19th Chief of Naval Operations, Zumwalt played a major role in U.S. military history, especially during the Vietnam War. A decorated war veteran, Zumwalt reformed U.S. Navy personnel policies in an effort to improve enlisted life and ease racial tensions. After he retired from a 32-year Navy career, he launched an unsuccessful campaign for the U.S. Senate.

2.3.1 Early life and education

Zumwalt was born in San Francisco, California, the son of Elmo Russell Zumwalt, M.D., and his wife, Frances (Frank) Zumwalt, M.D.,[1] both country doctors. Frances was raised Jewish, the daughter of Julius and Sarah Frank of Burlington, Vermont. Her family moved to Los Angeles, California, where she grew up. She became estranged from her parents for marrying outside the faith, as the Zumwalts were Christian.[2]

Zumwalt, an Eagle Scout and recipient of the Distinguished Eagle Scout Award from the Boy Scouts of America, attended Tulare Union High School in Tulare, California, where he became the valedictorian, and Rutherford Preparatory School in Long Beach, California.

2.3.2 Entrance into the US Navy

He had planned to become a doctor like his parents, but in 1939, Zumwalt was accepted to the United States Naval Academy at Annapolis, Maryland. As a midshipman at the USNA, he was president of the Trident Society, vice president of the Quarterdeck Society and the two-time winner of the June Week Public Speaking Contest (1940–41). Zumwalt also participated in intercollegiate debating and was a Company Commander (1941) and Regimental Three Striper (1942). He graduated with distinction and was commissioned as an ensign on June 19, 1942. He also received an honorary degree from Texas Tech University.

2.3.3 World War II

Zumwalt was assigned to USS *Phelps* (DD-360), a destroyer. In August 1943, *Phelps* was detached for instruction in the Operational Training Command-Pacific in San Francisco. In January 1944, Zumwalt reported for duty onboard USS *Robinson*. On this ship, he was awarded the Bronze Star with Valor device for *heroic service as Evaluator in the Combat Information Center ...in action against enemy Japanese battleships during the Battle for Leyte Gulf, October 25, 1944.*

After the end of World War II in August 1945, Zumwalt continued to serve until December 8, 1945, as the prize crew officer of the *Ataka*, a 1,200-ton Japanese river gunboat with a crew of 200. In this capacity, he took the first American-controlled ship since the outbreak of World War II up the Huangpu River to Shanghai, China. There, they helped to restore order and assisted in disarming the Japanese.

2.3.4   Command assignments

Zumwalt next served as executive officer of the destroyer USS *Saufley*, and in March 1946, was transferred to the destroyer USS *Zellars*, as Executive Officer and Navigator.

In January 1948, he was assigned to the Naval Reserve Officers Training Corps Unit of the University of North Carolina, where he remained until June 1950. That same month, he assumed command of USS *Tills*, a destroyer escort that was commissioned in a reserve status. The Tillis was placed in full active commission at Charleston Naval Shipyard on November 21, 1950, and he continued to command her until March 1951, when he joined the battleship USS *Wisconsin* as Navigator and served with the ship in operations in Korea.

Detached from USS *Wisconsin* in June 1952, he attended the Naval War College, Newport, Rhode Island, and in June 1953, he reported as Head of the Shore and Overseas Bases Section, Bureau of Naval Personnel, Navy Department, Washington, D.C. He also served as Officer and Enlisted Requirements Officer, and as Action Officer on Medicare Legislation. Completing that tour of duty in July 1955, he assumed command of the destroyer USS *Arnold J. Isbell*, participating in two deployments with the U.S. Seventh Fleet. In this assignment, he was commended by the Commander, Cruiser-Destroyer Forces, U.S. Pacific Fleet, for winning the Battle Efficiency Competition for his ship and for winning Excellence Awards in Engineering, Gunnery, Anti-Submarine Warfare, and Operations. In July 1957, he returned to the Bureau of Naval Personnel for further duty. In December 1957, he was transferred to the Office of the Assistant Secretary of the Navy (Personnel and Reserve Forces), and served as Special Assistant for Naval Personnel until November 1958, then as Special Assistant and Naval Aide until August 1959.

Ordered to the first ship built from the keel up as a guided missile frigate, USS *Dewey* (DLG-14), built at the Bath (Maine) Iron Works, he assumed command of that frigate at her commissioning in December 1959 and commanded her until June 1961. During this period of his command, Dewey earned the Excellence Award in Engineering, Supply, Weapons, and was runner-up in the Battle Efficiency Competition. He was a student at the National War College, Washington, D. C., during the 1961–1962 class year. In June 1962, he was assigned to the Office of the Assistant Secretary of Defense (International Security Affairs), Washington, D.C., where he served first as Desk Officer for France, Spain and Portugal, then as Director of Arms Control and Contingency Planning for Cuba. From December 1963 until June 21, 1965, he served as Executive Assistant and Senior Aide to the Honorable Paul H. Nitze, Secretary of the Navy. For duty in his tour in the offices of the Secretary of Defense and the Secretary of the Navy, he

was awarded the Legion of Merit.

2.3.5   Flag assignments

Vietnam

After his selection for the rank of Rear Admiral, Zumwalt assumed command of Cruiser-Destroyer Flotilla Seven on 24 July 1965 in San Diego.[3] He then served as Director, Systems Analysis Division, OPNAV (OP-96) from August 1966 to August 1968.[4] In September 1968, he became Commander Naval Forces, Vietnam, and Chief of the Naval Advisory Group, U.S. Military Assistance Command Vietnam (MACV) and was promoted to Vice Admiral in October 1968. Vice Admiral Zumwalt was the Navy adviser to General Creighton Abrams, Commander, MACV. Zumwalt always spoke very highly of Abrams, and said that Abrams was the most caring officer he had ever known.

Zumwalt's command was not a blue-water force, like the Seventh Fleet; it was a brown-water unit: he commanded the flotilla of Swift Boats that patrolled the coasts, harbors, and rivers of Vietnam. Among the swift-boat commanders were his son, Elmo Russell Zumwalt III, and later future Senator and U.S. Secretary of State John Kerry. Among his other forces were Task Force 115, the Coastal Surveillance Force, Task Force 116, the River Patrol Force, and Task Force 117, the joint Army-Navy Mobile Riverine Force.[5]

Chief of Naval Operations

President Richard Nixon nominated Zumwalt to be Chief of Naval Operations in April 1970. Upon being relieved as Commander Naval Forces, Vietnam, on May 15, 1970, he was awarded a second Navy Distinguished Service Medal for exceptionally meritorious service.

He assumed duties as Chief of Naval Operations and was promoted to full Admiral on July 1, 1970, and quickly began a series of moves intended to reduce racism and sexism in the Navy. These were disseminated in Navy-wide communications known as "Z-grams". These included orders authorizing beards (sideburns, mustaches, and longer groomed hair were also acceptable) and introducing beer-dispensing machines to barracks. Not all of these changes were well received by senior naval personnel. The measures to reduce discrimination against women and racial minorities were adamantly opposed by some.

Zumwalt instituted the 'Mod Squad' - Destroyer Squadron 26 and later 31 - to give promising young officers early command experience. Billets were a rank lower than normal.

Zumwalt reshaped the Navy's effort to replace large numbers of aging World War II-era vessels, a plan called "High-

Low." Instituted over the resistance of Admiral Hyman Rickover and others, High-Low sought to balance the purchase of high-end, nuclear-powered vessels with low-end, cheaper ones —- such as the Sea Control Ship—that could be bought in greater numbers. Rickover, the Father of the Nuclear Navy, preferred buying a few major ships to buying many ordinary ones. Zumwalt proposed four kinds of warships to fit the plan; in the end, only the *Pegasus* class of missile patrol boats and the *Oliver Hazard Perry* (FFG 7) class of guided missile frigates became reality, and only six out of the planned 100+ *Pegasus* class hydrofoils were built. But the *Perrys* stood as the most populous class of U.S. warships since World War II until the advent of the *Arleigh Burke* (DDG 51) destroyers. He was the last Chief of Naval Operations to live in Number One Observatory Circle before it became the official residence to the vice president. For Zumwalt, not pleased with the choice, this was reason enough to challenge Virginia Senator Harry F. Byrd, Jr. in the 1976 Senate election in Virginia.[6]

Elmo Zumwalt Jr. retired from the Navy on July 1, 1974, aged 53.

2.3.6 List of Z-grams

"Z-gram" was the semi-official title for policy directives issued by Elmo Zumwalt as Chief of Naval Operations (CNO). Many of these directives were efforts to reform outdated policies potentially contributing to difficulties recruiting and retaining qualified naval personnel during the period of United States withdrawal from the Vietnam War.

List of Z-grams

- Z-gram 1 (14 July 1970): convened a junior officer retention study group.

- Z-gram 2 (1 July 1970): Zumwalt's remarks upon taking office as CNO.

- Z-gram 3 (22 July 1970): Cryptographic procedures and Policy.

- Z-gram 4 (30 July 1970): authorized 30 days leave for officers with orders for a permanent change of station (PCS).

- Z-gram 5 (30 July 1970): instituted a test program aboard six ships to extend to 1st class petty officers the privilege of officers and chief petty officers (CPOs) to keep civilian clothing aboard ship for wearing on liberty.

- Z-gram 6 (11 August 1970): instituted a test program, funded entirely by deployed personnel to assist their families obtaining transportation and lodging to visit them in an overseas liberty port during holiday periods.

- Z-gram 7 (11 August 1970): directed commanding officers to assign sponsors for newly arriving personnel. The sponsors were normally of the same rank or rate and with similar marital and family status to assist the arriving family establishing themselves in the new location.

- Z-gram 8 (11 August 1970): extended the working hours of personnel writing officers' orders from 16:30 to 21:00 so those personnel would be available to answer telephone questions after duty hours of officers expecting orders.

- Z-gram 9 (14 August 1970): provided an alternative means of promotion to 1st class and CPO for highly motivated individuals who had five times failed the normal promotion examinations.

- Z-gram 10 (20 August 1970): required naval air stations to have an officer or CPO meet each arriving transient aircraft to coordinate aircraft servicing and assist flight crew with dining and temporary lodging.

- Z-gram 11 (24 August 1970): authorized continuing sea duty for enlisted men requesting it.

- Z-gram 12 (24 August 1970): authorized wearing of civilian clothes on shore bases during and after the evening meal by all enlisted personnel except recruits in basic training.

- Z-gram 13 (26 August 1970): directed commanding officers to grant 30 days of leave to at least half of their crew during the first 30 days following return from overseas deployment.

- Z-gram 14 (27 August 1970): abolished 18 collateral duties traditionally assigned to junior officers (including cigarette fund officer and cold weather officer) and encouraged assignment of another 18 collateral duties (including movie officer and athletics officer) to qualified senior petty officers.

- Z-gram 15 (28 August 1970): ordered all disbursing officers to provide all personnel with a statement of earnings prior to 30 October 1970 itemizing basic pay and allowances for clothing, quarters, sea duty, and hostile fire with taxes, deductions and allotments.

- Z-gram 16 (2 September 1970): established a computer database to assist enlisted personnel desiring a duty swap with a similarly qualified sailor on another ship or home port.

- Z-gram 17 (2 September 1970): raised the check-cashing limit at naval bases from $25 to $50.

- Z-gram 18 (4 September 1970): opened the Navy Finance Center around the clock to all disbursing officers processing urgent inquiries about pay and benefits.

- Z-gram 19 (4 September 1970): implemented an executive order from President Nixon to authorize an increased percentage of early promotions for officers.

- Z-gram 20 (8 September 1970): required all shore bases to provide washing facilities and lockers for enlisted personnel assigned dirty work in dungarees.

- Z-gram 21 (9 September 1970): encouraged commanding officers to provide compensatory time off for personnel standing watch on holidays.

- Z-gram 22 (9 September 1970): authorized shore bases to organize facility improvement teams for welfare, living and parking facilities.

- Z-gram 23 (12 September 1970): established the CPO advisory board to the CNO.

- Z-gram 24 (14 September 1970): established procedures for Navy wives to present complaints, viewpoints and suggestions to commanding officers of shore bases.

- Z-gram 25 (16 September 1970): authorized ships in port to reduce watch standing rotation from one day in four to one day in six.

- Z-gram 26 (21 September 1970): shifted responsibility for shore patrol staffing from shipboard to shore-based personnel at major naval bases.

- Z-gram 27 (21 September 1970): eliminated routine local operations over a weekend by ships sailing from their home port.

- Z-gram 28 (21 September 1970): was a status report on implementation of recommendations by retention study groups.

- Z-gram 29 (22 September 1970): encouraged commanding officers to allow leave for 5% of their crew during overseas deployments.

- Z-gram 30 (23 September 1970): established "hard-rock" officers' clubs for junior officers at five naval bases and encouraged other naval base officers' clubs to allow at least one room for casual dress, encourage unescorted young ladies to visit the clubs, and appoint younger officers to advise club managers about other measures to improve morale of junior officers.

- Z-gram 31 (23 September 1970): established a junior officer ship-handling competition whose winners would be able to pick their next duty assignment.

- Z-gram 32 (23 September 1970): allowed sailors to arrange their own re-enlistment ceremonies with assistance from their command.

- Z-gram 33 (25 September 1970): established a procedure to improve customer relations at naval Base Exchanges.

- Z-gram 34 (25 September 1970): eliminated the requirement for junior officers to own formal dinner dress uniforms.

- Z-gram 35 (25 September 1970): authorized alcoholic beverages in barracks and beer vending machines in senior enlisted barracks.

- Z-gram 36 (26 September 1970): encouraged commanding officers to improve the customer service ethic at base dispensaries and disbursing facilities.

- Z-gram 37 (26 September 1970): reduced the rank required for command of aviation squadrons from commander to lieutenant commander.

- Z-gram 38 (28 September 1970): instructed commanding officers to eliminate scheduling of work routine on Sundays and holidays unless ship is deployed overseas.

- Z-gram 39 (5 October 1970): extended the operating hours of 25 large base commissaries to reduce crowds on Saturday mornings and paydays.

- Z-gram 40 (7 October 1970): gave sailors the option of being paid either in cash or by check.

- Z-gram 41 (21 October 1970): established a Command Excellence chair at the Naval War College to be filled by a commander or captain with a record of outstanding performance in command.

- Z-gram 42 (13 October 1970): allowed junior officers to request sea duty as their first choice for initial duty assignment.

- Z-gram 43 (13 October 1970): encouraged commanding officers to help disbursing officers speedily process large travel reimbursement claims.

- Z-gram 44 (13 October 1970): encouraged assignment of senior petty officers to stand in-port officer of the deck watches to reduce junior officer workload.

- Z-gram 45 (15 October 1970): encouraged commanding officers to increase support services to families of prisoners of war.

- Z-gram 46 (15 October 1970): reduced routine paperwork required for the 3M planned maintenance system inspections and documentation.

- Z-gram 47 (20 October 1970): increased responsibilities of department heads and executive officers of ships being deactivated.

- Z-gram 48 (23 October 1970): established a new Bureau of Naval Personnel office focused on providing information to dependent families of active duty personnel.

- Z-gram 49 (23 October 1970): required half of personnel on awards boards to be below the rank of commander.

- Z-gram 50 (23 October 1970): encouraged ships returning from overseas deployments to use shore based utilities to allow leave for increased numbers of engineering personnel.

- Z-gram 51 (23 October 1970): established a uniform breast insignia for officers in charge of brown-water navy boats.

- Z-gram 52 (23 October 1970): (CLASSIFIED)

- Z-gram 53 (2 November 1970): authorized annual publication of a list of job assignments available to junior officers, emphasizing geographical locations and required qualifications for career planning.

- Z-gram 54 (2 November 1970): outlined procedures for junior personnel to make suggestions to CNO.

- Z-gram 55 (4 November 1970): established pilot program for improving Navy human resources management.

- Z-gram 56 (9 November 1970): established a program similar to Z-16 for officers desiring a duty swap with a similarly qualified officer on another ship or home port.

- Z-gram 57 (10 November 1970): eliminated a broad spectrum of selectively enforced regulations and specified relaxed interpretations of others related to grooming standards and wearing of uniforms, so the vast majority of sailors would not be penalized by policies designed to constrain a few abusing the trust and confidence of less stringent rules.

- Z-gram 58 (14 November 1970): required ships' stores afloat to accept checks in payment for purchases.

- Z-gram 59 (14 November 1970): established a program for officers to spend a year of independent research and study for professional development in areas mutually beneficial to the officer and the Navy.

- Z-gram 60 (18 November 1970): encouraged all major naval installations to install a recording answering device on one telephone to receive suggestions.

- Z-gram 61 (19 November 1970): Authorized warrant officers and senior petty officers afloat to serve as communications watch officers and registered publications custodians.

- Z-gram 62 (27 November 1970): established a Naval War College forum to discuss improved naval personnel policies and present their views to CNO and Secretary of the Navy.

- Z-gram 63 (30 November 1970): reduced by 25% the number of publications to be maintained by ships.

- Z-gram 64 (3 December 1970): encouraged commanding officers to increase the opportunities for junior officers to practice ship handling.

- Z-gram 65 (5 December 1970): listed incentives for officers to volunteer for duty in Vietnam.

- Z-gram 66 (17 December 1970): directed every navy facility to appoint a minority group officer or senior petty officer as a minority affairs assistant to the commanding officer.

- Z-gram 67 (22 December 1970): streamlined required inspection procedures to reduce the amount of time required for preparation and execution.

- Z-gram 68 (23 December 1970): expanded the civilian clothing privilege explored in Z-gram 5 to all petty officers on all ships.

- Z-gram 69 (28 December 1970): eliminated command of a deep draft ship from the requirements for promotion to admiral.

- Z-gram 70 (21 January 1971): clarified grooming standards and working uniform regulations addressed by Z-gram 57 to reflect contemporary hair styles and allow wearing working uniforms while commuting between the base and off-base housing.[7]

2.3.7 Later years

In 1976, he unsuccessfully ran as a Democratic candidate for the United States Senate from Virginia, and was defeated by incumbent senator Harry F. Byrd, Jr.. Later, he held the presidency of the American Medical Building Corporation in Milwaukee, Wisconsin.

2.3.8 Family and Agent Orange controversy

While serving in Shanghai in 1945, Zumwalt met and married Mouza Coutelais-du-Roche, whose French-Russian family was living there. She returned with him to the United States. They had four children: Elmo Russell Zumwalt III, James Gregory Zumwalt ; Ann F. Zumwalt Coppola and Mouzetta C. Zumwalt-Weathers.

Zumwalt's eldest son, Elmo Zumwalt III, served as lieutenant on one of Zumwalt's patrol boats during the Vietnam War. In January 1983, he was diagnosed with lymphoma, and in 1985 it was discovered that he also had Hodgkin's disease.[8] In addition, his son Elmo Russell Zumwalt IV had been born in 1977 with severe learning disabilities. Admiral Zumwalt and his family were convinced that both son and grandson were victims of Agent Orange, which the admiral had ordered to be sprayed over the Mekong Delta to kill vegetation and drive "the Viet Cong back 1,000 yards off the water's edge."[8]

In an article published in The New York Times in 1986, Elmo Zumwalt III said:

> I am a lawyer and I don't think I could prove in court, by the weight of the existing scientific evidence, that Agent Orange is the cause of all the medical problems - nervous disorders, cancer and skin problems - reported by Vietnam veterans, or of their children's severe birth defects. But I am convinced that it is.[8]

Admiral Zumwalt and his son collaborated with writer John Pekkanen to create the book *My Father, My Son*, published by MacMillan in September 1986, where they discussed the family tragedy of his son's battle with cancer. In 1988, the book was made into a TV movie with the same name, starring Karl Malden as the admiral and Keith Carradine as his son.[9]

Elmo Zumwalt III died from his cancer on August 14, 1988, at the age of 42,[8] three months after the TV movie was shown.[9]

During his son's illness in the early 1980s, Admiral Zumwalt was very active in lobbying Congress to establish a national registry of bone marrow donors. Such donors serve patients who do not have suitably matched bone marrow donors in their families. This was ultimately a disinterested act, since his son was able to receive a transplant from his own sister, but many patients don't have close relatives who are able and willing to help in this heroic way. His efforts were a major factor in the founding of the National Marrow Donor Program (NMDP) in July 1986. Admiral Zumwalt was the first chairman of the NMDP's Board of Directors.

In his later years, Elmo Zumwalt, Jr. resided in Arlington County, Virginia.

2.3.9 Books

After he retired, Admiral Zumwalt wrote *On Watch: a Memoir,* published by Quadrangle Books in 1976. It reviews his Navy career and includes reprints of all the Z-Grams he issued as CNO.

2.3.10 Death

Zumwalt's casket being carried by pallbearers at his funeral in January 2000.

Zumwalt died on January 2, 2000, aged 79, at the Duke University Medical Center in Durham, North Carolina, from a rare form of lung cancer called pleural mesothelioma. Most likely, at some time in his naval career, Zumwalt was exposed to asbestos, which was widely used on naval vessels until it was banned during the 1980s after its hazards became widely known.

His funeral service was held at the Naval Academy Chapel.

In his eulogy President Bill Clinton called Zumwalt "the conscience of the United States Navy."[10]

2.3.11 Legacy

The United States Navy's DD(X) guided missile destroyer program has been named the Zumwalt class in his honor, and its lead ship will bear his name USS *Zumwalt* by Navy tradition.

In 2013, the Mesothelioma Center for Excellence at the VA West Los Angeles Medical Center was renamed the Elmo Zumwalt Treatment & Research Center specializing in mesothelioma research, particularly for veterans who may have been exposed to asbestos during their service.

2.3.12 Dates of rank

United States Naval Academy Midshipman – Class of 1942

2.3.13 Awards and decorations

U.S. military awards and decorations

U.S. civilian awards

Foreign awards

Foreign unit awards

Boy Scouts of America awards

2.3.14 Miscellaneous

- Zumwalt's picture hangs in the War Remnants Museum in Ho Chi Minh City, near pictures of John Kerry, Robert McNamara, Warren Christopher, and other American dignitaries, in commemoration of a visit he made after normalization of relations between Vietnam and the United States.[11]

- In his first book, *On Watch*, Zumwalt quoted at length an interview with Admiral Hyman G. Rickover, regarded as the Father of the Nuclear Navy and who interviewed all officers with responsibilities involving nuclear propulsion. Rickover and Zumwalt had a combative conversation, with Zumwalt referring to it as a humbling experience.

- Zumwalt was a member of Sigma Phi Epsilon fraternity. He was initiated in 1980.

- In 1994 the U.S. Navy Memorial Foundation awarded Zumwalt its Lone Sailor Award for his distinguished naval career.

2.3.15 References

[1] *Contemporary Authors Online*, Gale, 2009. Reproduced in Biography Resource Center. Farmington Hills, Michigan: Gale, 2009.

[2] Berman, Larry (2012). *Zumwalt: The Life and Times of Admiral Elmo Russell "Bud" Zumwalt, Jr.* HarperCollins.

[3] Berman, Zumwalt, 154

[4] Berman, Zumwalt, 433.

[5] Berman, Zumwalt, 171

[6] Bush, B. (2010). *Barbara Bush: A Memoir*. Scribner. ISBN 9781451603958.

[7] Zumwalt, Elmo (1971). "Z-grams". *Proceedings* (United States Naval Institute) **97** (819): 293–298.

[8] New York Times, August 14, 1988: *Elmo R. Zumwalt 3d, 42, Is Dead; Father Ordered Agent Orange Use* Linked 2014-09-07

[9] IMDb: *My Father, My Son (1988)* Linked 2014-09-07

[10] "William J. Clinton: "Remarks at Funeral Services for Elmo R. Zumwalt, Jr., in Annapolis, Maryland," January 10, 2000". The American Presidency Project. Retrieved 2013-11-28.

[11]

2.3.16 Further reading

- *My Father, My Son* by Elmo R. Zumwalt, Jr. and Elmo R. Zumwalt III, with John Pekkanen. (Dell Publishing Company, ISBN 0-440-15973-3)

- *On Watch: a memoir* by Elmo R. Zumwalt, Jr. (The New York Times Book Co., ISBN 0-8129-0520-2)

- *Admiral Elmo R. Zumwalt Jr., Texas Tech University Series* Texas Tech University's Virtual Vietnam archive

2.3.17 External links

- "Elmo Zumwalt". Find a Grave. Retrieved 2008-07-13.

- Z-grams: A List of Policy Directives Issued by Admiral Zumwalt 1 July 1970 to 1 July 1974

- 1972 Time magazine article on resistance to Zumwalt's policies, "Keelhauling the Navy"

- Zumwalt, ADM Elmo R., Jr., U.S. Navy (Ret.)

- Zumwalt Staff Officers, Volume I

- "Me and "Z" – Admiral Elmo R. Zumwalt | Fox News". foxnews.com. Retrieved 2014-07-12.

2.4 Free-electron laser

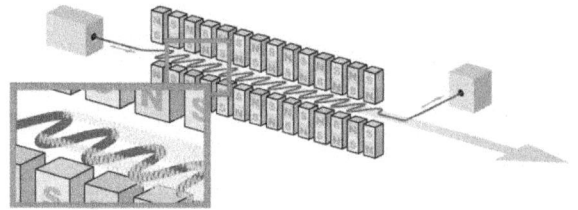

Schematic representation of an undulator, at the core of a free-electron laser.

2.4.1 Beam creation

The undulator of FELIX.

The free-electron laser FELIX at the FOM Institute for Plasma Physics Rijnhuizen (nl), Nieuwegein, The Netherlands.

A **free-electron laser** (FEL), is a type of laser whose lasing medium consists of very-high-speed electrons moving freely through a magnetic structure,[1] hence the term *free electron*.[2] The free-electron laser is tunable and has the widest frequency range of any laser type,[3] currently ranging in wavelength from microwaves, through terahertz radiation and infrared, to the visible spectrum, ultraviolet, and X-ray.[4]

The term free-electron lasers was coined by John Madey in 1976 at Stanford University.[5] The work emanates from research done by Hans Motz and his coworkers, who built an undulator at Stanford in 1953,[6][7] using the wiggler magnetic configuration which is the heart of a free electron laser. Madey used a 43-MeV electron beam[8] and 5 m long wiggler to amplify a signal.

To create a FEL, a beam of electrons is accelerated to almost the speed of light. The beam passes through an undulator, a side to side magnetic field produced by a periodic arrangement of magnets with alternating poles across the beam path. The direction of the beam is called the longitudinal direction, while the direction across the beam path is called transverse. This array of magnets is called an undulator or a wiggler, because it forces the electrons in the beam to wiggle transversely along a sinusoidal path about the axis of the undulator.

The transverse acceleration of the electrons across this path results in the release of photons (synchrotron radiation), which are monochromatic but still incoherent, because the electromagnetic waves from randomly distributed electrons interfere constructively and destructively in time, and the resulting radiation power scales linearly with the number of electrons. If an external laser is provided or if the synchrotron radiation becomes sufficiently strong, the transverse electric field of the radiation beam interacts with the transverse electron current created by the sinusoidal wiggling motion, causing some electrons to gain and others to lose energy to the optical field via the ponderomotive force.

This energy modulation evolves into electron density (current) modulations with a period of one optical wavelength.

The electrons are thus clumped, called *microbunches*, separated by one optical wavelength along the axis. Whereas conventional undulators would cause the electrons to radiate independently, the radiation emitted by the bunched electrons are in phase, and the fields add together coherently.

The FEL radiation intensity grows, causing additional microbunching of the electrons, which continue to radiate in phase with each other.[9] This process continues until the electrons are completely microbunched and the radiation reaches a saturated power several orders of magnitude higher than that of the undulator radiation.

The wavelength of the radiation emitted can be readily tuned by adjusting the energy of the electron beam or the magnetic-field strength of the undulators.

FELs are relativistic machines. The wavelength of the emitted radiation, λ_r, is given by [10]

$$\lambda_r = \frac{\lambda_u}{2\gamma^2}(1 + K^2)$$

or when the wiggler strength parameter K, discussed below, is small

$$\lambda_r \propto \frac{\lambda_u}{2\gamma^2}$$

where λ_u is the undulator wavelength (the spatial period of the magnetic field), γ is the relativistic Lorentz factor and the proportionality constant depends on the undulator geometry and is of the order of 1.

This formula can be understood as a combination of two relativistic effects. Imagine you are sitting on an electron passing through the undulator. Due to Lorentz contraction the undulator is shortened by a γ factor and the electron experiences much shorter undulator wavelength λ_u/γ. However, the radiation emitted at this wavelength is observed in the laboratory frame of reference and the relativistic Doppler effect brings the second γ factor to the above formula. Rigorous derivation from Maxwell's equations gives the divisor of 2 and the proportionality constant. In an X-ray FEL the typical undulator wavelength of 1 cm is transformed to X-ray wavelengths on the order of 1 nm by $\gamma \approx 2000$, i.e. the electrons have to travel with the speed of $0.9999998c$.

Wiggler strength parameter K

K, a dimensionless parameter, tells the wiggler strength as the relationship between the length of a period and the radius of bend,[11]

$$K = \frac{\gamma\lambda_u}{2\pi\rho} = \frac{eB_0\lambda_u}{\sqrt{8}\pi m_e c}$$

where ρ is the bending radius, B_0 is the applied magnetic field and m_e the electron mass.

Quantum effects

In most cases, the theory of classical electromagnetism adequately accounts for the behavior of free electron lasers.[12] For sufficiently short wavelengths, quantum effects of electron recoil and shot noise may have to be considered.[13]

2.4.2 Large facilities required

Free-electron lasers require the use of an electron accelerator with its associated shielding, as accelerated electrons can be a radiation hazard if not properly contained. These accelerators are typically powered by klystrons, which require a high voltage supply. The electron beam must be maintained in a vacuum which requires the use of numerous vacuum pumps along the beam path. While this equipment is bulky and expensive, free-electron lasers can achieve very high peak powers, and the tunability of FELs makes them highly desirable in many disciplines, including chemistry, structure determination of molecules in biology, medical diagnosis, and nondestructive testing.

2.4.3 X-ray laser without mirrors

The lack of a material to make mirrors that can reflect extreme ultraviolet and x-rays means that FELs at these frequencies cannot use a resonant cavity like other lasers, which reflects the radiation so it makes multiple passes through the undulator. Consequently, in an X-ray FEL the output beam is produced by a single pass of radiation through the undulator; there must be enough amplification over a single pass to produce an adequately bright beam.

X-ray free electron lasers use long undulators. The underlying principle of the intense pulses from the X-ray laser lies in the principle of self-amplified spontaneous emission (SASE), which leads to the microbunching. Initially all electrons are distributed evenly and they emit incoherent spontaneous radiation only. Through the interaction of this radiation and the electrons' oscillations, they drift into microbunches separated by a distance equal to one radiation wavelength. Through this interaction, all electrons begin emitting coherent radiation in phase. All emitted radiation can reinforce itself perfectly whereby wave crests and wave troughs are always superimposed on one another in the best possible way. This results in an exponential increase of emitted radiation power, leading to high beam intensities and laser-like properties.[14] Examples of facilities operating on the SASE FEL principle include the Free electron LASer in Hamburg (FLASH), the Linac Coherent Light Source (LCLS) at the SLAC National Accelerator Laboratory, the European x-ray free electron laser (XFEL) in Hamburg, the SPring-8 Compact SASE Source (SCSS), the SwissFEL at the Paul Scherrer Institute (Switzerland) and, as of 2011, the SACLA at the RIKEN Harima Institute in Japan.

2.4.4 Self seeding

One problem with SASE FELs is the lack of temporal coherence due to a noisy startup process. To avoid this, one can "seed" an FEL with a laser tuned to the resonance of the FEL. Such a temporally coherent seed can be produced by more conventional means, such as by high-harmonic generation (HHG) using an optical laser pulse. This results in coherent amplification of the input signal; in effect, the output laser quality is characterized by the seed. While HHG seeds are available at wavelengths down to the extreme ultraviolet, seeding is not feasible at x-ray wavelengths due to the lack of conventional x-ray lasers. In late 2010, in Italy, the seeded-FEL source FERMI@Elettra [15] started commissioning, at the Sincrotrone Trieste Laboratory. FERMI@Elettra is a single-pass FEL user-facility covering the wavelength range from 100 nm (12 eV) to 10 nm (124 eV), located next to the third-generation synchrotron radiation facility ELETTRA in Trieste, Italy. The advent of femtosecond lasers has revolutionized many areas of science from solid state physics to biology.

In 2012, scientists working on the LCLS overcame the seeding limitation for x-ray wavelengths by self-seeding the laser with its own beam after being filtered through a diamond monochromator. The resulting intensity and monochromaticity of the beam were unprecedented and allowed new experiments to be conducted involving manipulating atoms and imaging molecules. Other labs around the world are incorporating the technique into their equipment.[16][17]

2.4.5 Applications

Medical

Surgery   Research by Glenn Edwards and colleagues at Vanderbilt University's FEL Center in 1994 found that soft tissues including skin, cornea, and brain tissue could be cut, or ablated, using infrared FEL wavelengths around 6.45 micrometres with minimal collateral damage to adjacent

tissue.[18][19] This led to surgeries on humans, the first ever using a free-electron laser. Starting in 1999, Copeland and Konrad performed three surgeries in which they resected meningioma brain tumors.[20] Beginning in 2000, Joos and Mawn performed five surgeries that cut a window in the sheath of the optic nerve, to test the efficacy for optic nerve sheath fenestration.[21] These eight surgeries produced results consistent with the standard of care and with the added benefit of minimal collateral damage. A review of FELs for medical uses is given in the 1st edition of Tunable Laser Applications.[22]

Fat removal   Several small, clinical lasers tunable in the 6 to 7 micrometre range with pulse structure and energy to give minimal collateral damage in soft tissue were created. At Vanderbilt, there exists a Raman shifted system pumped by an Alexandrite laser.[23]

Rox Anderson proposed the medical application of the free-electron laser in melting fats without harming the overlying skin.[24] At infrared wavelengths, water in tissue was heated by the laser, but at wavelengths corresponding to 915, 1210 and 1720 nm, subsurface lipids were differentially heated more strongly than water. The possible applications of this selective photothermolysis (heating tissues using light) include the selective destruction of sebum lipids to treat acne, as well as targeting other lipids associated with cellulite and body fat as well as fatty plaques that form in arteries which can help treat atherosclerosis and heart disease.[25]

Biology

Exceptionally bright and fast X-rays can image proteins using a sheet just one molecule thick. This technique allows first-time imaging of proteins that do not stack in a way that allows imaging by conventional techniques, 25% of the total number of proteins. Resolutions of 0.8 nm have been achieved with pulse durations of 30 femtoseconds. To get a clear view, a resolution of 0.1–0.3 nm is required. The short pulse durations prevented the lasers from destroying the molecules. The bright, fast X-rays were produced at the Linac Coherent Light Source at SLAC. As of 2014 LCLS was the world's most powerful X-ray FEL.[26]

Military

FEL technology is being evaluated by the US Navy as a candidate for an antiaircraft and missile directed-energy weapon. The Thomas Jefferson National Accelerator Facility's FEL has demonstrated over 14 kW power output.[27] Compact multi-megawatt class FEL weapons are undergoing research.[28] On June 9, 2009 the Office of Naval Research announced it had awarded Raytheon a contract to develop a 100 kW experimental FEL.[29] On March 18, 2010 Boeing Directed Energy Systems announced the completion of an initial design for U.S. Naval use.[30] A prototype FEL system was demonstrated, with a full-power prototype scheduled by 2018.[31]

2.4.6 See also

- Bremsstrahlung

- Cyclotron radiation

- Electron wake

- Gyrotron

- International Linear Collider

- Synchrotron radiation

2.4.7 References

[1] Huang, Z.; Kim, K. J. (2007). "Review of x-ray free-electron laser theory". *Physical Review Special Topics - Accelerators and Beams* **10** (3). doi:10.1103/PhysRevSTAB.10.034801.

[2] "Southeastern Universities Research Association Thomas Jefferson National Accelerator Facility". Retrieved 2015-11-19.

[3] F. J. Duarte (Ed.), *Tunable Lasers Handbook* (Academic, New York, 1995) Chapter 9.

[4] "New Era of Research Begins as World's First Hard X-ray Laser Achieves "First Light"". SLAC National Accelerator Laboratory. April 21, 2009. Retrieved 2013-11-06.

[5] Hans Motz, W. Thon, R.N. Whitehurst, Experiments on radiation by fast electron beams, *Journal of Applied Physics*, 24(7):826-833, 1953.

[6] Motz, Hans (1951). "Applications of the Radiation from Fast Electron Beams". *Journal of Applied Physics* **22** (5): 527. doi:10.1063/1.1700002.

[7] Motz, H.; Thon, W.; Whitehurst, R. N. (1953). "Experiments on Radiation by Fast Electron Beams". *Journal of Applied Physics* **24** (7): 826. doi:10.1063/1.1721389.

[8] "Phys. Rev. Lett. 38, 892 (1977): First Operation of a Free-Electron Laser". Prl.aps.org. Retrieved 2014-02-17.

[9] Feldhaus, J.; Arthur, J.; Hastings, J. B. (2005). "X-ray free-electron lasers". *Journal of Physics B: Atomic, Molecular and Optical Physics* **38** (9): S799. doi:10.1088/0953-4075/38/9/023.

[10] Neil, G.; Merminga, L. (2002). "Technical approaches for high-average-power free-electron lasers". *Reviews of Modern Physics* **74** (3): 685. doi:10.1103/RevModPhys.74.685.

[11] Robert Soliday (2006-09-05). "WIGGLER". Argon National laboratory.

[12] Fain, B.; Milonni, P. W. (1987). "Classical stimulated emission". *Journal of the Optical Society of America B* **4**: 78. doi:10.1364/JOSAB.4.000078.

[13] Benson, S.; Madey, J. M. J. (1984). "Quantum fluctuations in XUV free electron lasers". *AIP Conference Proceedings* **118**. p. 173. doi:10.1063/1.34633.

[14] "XFEL information webpages". Retrieved 2007-12-21.

[15] "FERMI / HomePage". Elettra.trieste.it. 2013-10-24. Retrieved 2014-02-17.

[16] Amann, J.; Berg, W.; Blank, V.; Decker, F. -J.; Ding, Y.; Emma, P.; Feng, Y.; Frisch, J.; Fritz, D.; Hastings, J.; Huang, Z.; Krzywinski, J.; Lindberg, R.; Loos, H.; Lutman, A.; Nuhn, H. -D.; Ratner, D.; Rzepiela, J.; Shu, D.; Shvyd'ko, Y.; Spampinati, S.; Stoupin, S.; Terentyev, S.; Trakhtenberg, E.; Walz, D.; Welch, J.; Wu, J.; Zholents, A.; Zhu, D. (2012). "Demonstration of self-seeding in a hard-X-ray free-electron laser". *Nature Photonics* **6** (10): 693. doi:10.1038/nphoton.2012.180.

[17] ""Self-seeding" promises to speed discoveries, add new scientific capabilities". SLAC National Accelerator Laboratory. August 13, 2012. Retrieved 2013-11-06.

[18] Edwards, G.; Logan, R.; Copeland, M.; Reinisch, L.; Davidson, J.; Johnson, B.; MacIunas, R.; Mendenhall, M.; Ossoff, R.; Tribble, J.; Werkhaven, J.; O'Day, D. (1994). "Tissue ablation by a free-electron laser tuned to the amide II band". *Nature* **371** (6496): 416. doi:10.1038/371416a0.

[19] "Laser light from Free-Electron Laser used for first time in human surgery". Retrieved 2010-11-06.

[20] Glenn S. Edwards et al., Rev. Sci. Instrum. 74 (2003) 3207

[21] MacKanos, M. A.; Joos, K. M.; Kozub, J. A.; Jansen, E. D. (2005). "Corneal ablation using the pulse stretched free electron laser". *Ophthalmic Technologies XV*. Ophthalmic Technologies XV **5688**. p. 177. doi:10.1117/12.596603.

[22] F. J. Duarte (12 December 2010). "6". *Tunable Laser Applications, Second Edition*. CRC Press. ISBN 978-1-4200-6058-4.

[23] "Efficiency and Plume Dynamics for Mid-IR Laser Ablation of Cornea". 2009-03-18. Retrieved 2010-11-06.

[24] "BBC health". *BBC News*. 2006-04-10. Retrieved 2007-12-21.

[25] "Dr Rox Anderson treatment". Retrieved 2007-12-21.

[26] "Super-bright, fast X-ray free-electron lasers can now image single layer of proteins". KurzweilAI. doi:10.1107/S2052252514001444. Retrieved 2014-02-17.

[27] "Jefferson Lab FEL". Retrieved 2009-06-08.

[28] "Airborne megawatt class free-electron laser for defense and security". Retrieved 2007-12-21.

[29] "Raytheon Awarded Contract for Office of Naval Research's Free Electron Laser Program". Retrieved 2009-06-12.

[30] "Boeing Completes Preliminary Design of Free Electron Laser Weapon System". Retrieved 2010-03-29.

[31] "Breakthrough Laser Could Revolutionize Navy's Weaponry". Fox News. 2011-01-20. Retrieved 2011-01-22.

2.4.8 Further reading

- Madey, John, "Stimulated emission of bremsstrahlung in a periodic magnetic field". J. Appl. Phys. 42, 1906 (1971)

- Madey, John, Stimulated emission of radiation in periodically deflected electron beam, US Patent 38 22 410,1974

- Boscolo, et al., "*Free-Electron Lasers and Masers on Curved Paths*". Appl. Phys., (Germany), vol. 19, No. 1, pp. 46–51, May 1979.

- Deacon et al., "*First Operation of a Free-Electron Laser*". Phys. Rev. Lett., vol. 38, No. 16, Apr. 1977, pp. 892–894.

- Elias, et al., "*Observation of Stimulated Emission of Radiation by Relativistic Electrons in a Spatially Periodic Transverse Magnetic Field*", Phys. Rev. Lett., 36 (13), 1976, p. 717.

- Gover, "*Operation Regimes of Cerenkov-Smith-Purcell Free Electron Lasers and T. W. Amplifiers*". Optics Communications, vol. 26, No. 3, Sep. 1978, pp. 375–379.

- Gover, "*Collective and Single Electron Interactions of Electron Beams with Electromagnetic Waves and Free Electrons Lasers*". App. Phys. 16 (1978), p. 121.

- "*The FEL Program at Jefferson Lab*"

- Brau, Charles (1990). "Free-Electron Lasers". Boston: Academic Press, Inc.

- Paolo Luchini, Hans Motz, *Undulators and Free-electron Lasers*, Oxford University Press, 1990.

2.4.9 External links

- Lightsources.org

- FERMI, the new FEL at the ELETTRA synchrotron in Triest

- Free-Electron Laser Open Book (National Academies Press)

- The World Wide Web Virtual Library: Free-Electron Laser research and applications

- European XFEL

- PSI SwissFEL

- SPring-8 Compact SASE Source

- Electron beam transport system and diagnostics of the Dresden FEL

- The Free Electron Laser for Infrared eXperiments FE-LIX

- W. M. Keck Free Electron Laser Center

- Jefferson Lab's Free-Electron Laser Program

- Free-Electron Lasers: The Next Generation by Davide Castelvecchi New Scientist, January 21, 2006

- Airborne megawatt class free-electron laser for defense and security

- FERMI@Elettra Free-Electron Laser Project

- Center for Free-Electron Laser Science (CFEL)

2.5 Guided missile destroyer

A **guided-missile destroyer** is a destroyer designed to launch guided missiles. Many are also equipped to carry out anti-submarine, anti-air, and anti-surface operations. The NATO standard designation for these vessels is **DDG**. Nations vary in their use of destroyer **D** designation their hull pennant numbering, either prefixing, or dropping it altogether. The U.S. Navy has adopted the classification **DDG** in the American hull classification system.

In addition to the guns that destroyers have, a guided-missile destroyer is usually equipped with two large missile magazines, usually in vertical-launch cells. Some guided-missile destroyers contain powerful radar systems, such as the United States' Aegis Combat System, and may be adopted for use in an anti-missile or ballistic-missile defense role. This is especially true of navies that no longer operate cruisers, as other vessels must be adopted to fill in the gap.

The Japanese guided missile destroyer JDS Kongō *(DDG-173) firing a Standard Missile 3 anti-ballistic missile.*

2.5.1 Active and planned guided missile destroyers

Royal Australian Navy

- *Hobart*-class destroyer

    - HMAS *Hobart* (Under construction – D39?)
    - HMAS *Brisbane* (Under construction – D41?)
    - HMAS *Sydney* (Under construction)

HMCS Iroquois *(DDG-280), an* Iroquois-*class destroyer*

Royal Canadian Navy

- *Iroquois*-class destroyer

    - HMCS *Athabaskan* (DDH 282)

People's Liberation Army Navy of China

Kunming *(D172), a Chinese Type 052D destroyer*

- Type 055 destroyer (under construction)
- Type 052D destroyer

    - *Kunming* (172)
    - *Changsha* (173)
    - *Hefei* (174)
    - *Yinchuan* (175) (Sea trial)
    - (117) (Sea trial)
    - (118) (Fitting out)
    - (119) (Fitting out)
    - (120) (Fitting out)
    - (154) (Under construction)
    - (155) (Under construction)

- Type 052C (Luyang II class) destroyer

    - *Lanzhou* (170)
    - *Haikou* (171)
    - *Changchun* (150)
    - *Zhengzhou* (151)
    - *Jinan* (152)
    - Xi'an (153)

- Type 052B (Luyang I class) destroyer

    - *Guangzhou* (168)
    - *Wuhan* (169)

- Type 052 (Luhu class) destroyer

- *Harbin* (112)
- *Qingdao* (113)
- Type 051C (Luzhou class) destroyer
  - *Shenyang* (115)
  - *Shijiazhuang* (116)
- Type 051B (Luhai class) destroyer
  - *Shenzhen* (167)
- Type 051 (Luda class) destroyer
  - *Xining* (108)
  - *Kaifeng* (109)
  - *Dalian* (110)
  - *Chongqing* (133)
  - *Zunyi* (134)
  - *Nanchang* (163)
  - *Guilin* (164)
  - *Zhanjiang* (165)
  - *Zhuhai* (166)
- *Sovremenny*-class destroyer
  - *Hangzhou* (136)
  - *Fuzhou* (137)
  - *Taizhou* (138)
  - *Ningbo* (139)

French Navy

Forbin (D620), a Horizon-class destroyer

Although the French Navy no longer uses the term "destroyer" (French: destructeur), the largest frigates are assigned pennant numbers with flag superior "D", which designates destroyer.

- Horizon-class frigate
  - *Forbin* (D620)
  - *Chevalier Paul* (D621)
- *Cassard*-class frigate
  - *Cassard* (D 614)
  - *Jean Bart* (D615)
- *Georges Leygues*-class frigate
  - *Dupleix* (D641)
  - *Montcalm* (D642)
  - *Jean de Vienne* (D643)
  - *Primauguet* (D644)
  - *La Motte-Picquet* (D645)
  - *Latouche-Tréville* (D646)
- *Aquitaine*-class frigate

Indian Navy

INS Kolkata *(D63), a* Kolkata-*class destroyer*

- *Visakhapatnam*-class destroyer (Under construction)
  - INS *Visakhapatnam* (D66) [1]
  - INS *Porbandar* (D67)
  - INS *Marmagoa* (D68) [2]
  - INS Paradip(D69)
- *Kolkata*-class destroyer
  - INS *Kolkata*
  - INS *Kochi*
  - INS *Chennai*

INS Mumbai, *a* Delhi-*class Destroyer*

Delhi-class destroyer

- INS *Delhi*
- INS *Mysore*
- INS *Mumbai*

- *Rajput*-class destroyer

  - INS *Rajput*
  - INS *Rana*
  - INS *Ranjit*
  - INS *Ranvir*
  - INS *Ranvijay*

Italian Navy

Caio Duilio *(D554), an Orizzonte-class destroyer*

- *Durand de la Penne*-class destroyer
  - *Luigi Durand De La Penne* (D560)
  - *Francesco Mimbelli* (D561)
- Orizzonte-class destroyer
  - *Andrea Doria* (D553)
  - *Caio Duilio* (D554)

Japan Maritime Self-Defense Force

JDS Ashigara *(DDG-178), an* Atago-*class destroyer*

- *Amatsukaze*-class destroyer
  - JDS *Amatsukaze* (DDG-163)
- *Tachikaze*-class destroyer
  - JDS *Tachikaze* (DDG-168)
  - JDS *Asakaze* (DDG-169)
  - JDS *Sawakaze* (DDG-170)
- *Hatakaze*-class destroyer
  - JDS *Hatakaze* (DDG-171)
  - JDS *Shimakaze* (DDG-172)
- *Kongō*-class destroyer
  - JDS *Kongō* (DDG-173)
  - JDS *Kirishima* (DDG-174)
  - JDS *Myōkō* (DDG-175)
  - JDS *Chōkai* (DDG-176)
- *Atago*-class destroyer
  - JDS *Atago* (DDG-177)
  - JDS *Ashigara* (DDG-178)

Russian Navy

- Kashin-class destroyer
  - *Smetlivy* (810)
- *Sovremenny*-class destroyer
  - *Bystryy* (715)
  - *Gremyashchiy* (406)
  - *Bespokoynyy* (620)

Nastoychivyy *(610), a* Sovremenny-*class destroyer*

- *Nastoychivyy* (610)
- *Admiral Ushakov* (434)

- *Udaloy*-class destroyer

  - *Vice-Admiral Kulakov*
  - *Admiral Tributs* (552)
  - *Marshal Shaposhnikov* (543)
  - *Severomorsk* (619)
  - *Admiral Levchenko* (605)
  - *Admiral Vinogradov* (572)
  - *Admiral Panteleyev* (548)
  - *Admiral Chabanenko* (650)

ROCS Tso Ying *(DDG-1803)*

Republic of China Navy (Taiwan)

- *Kee Lung*-class destroyer (ex-*Kidd* class)

  - ROCS *Kee Lung* (DDG-1801)
  - ROCS *Su Ao* (DDG-1802)
  - ROCS *Tso Ying* (DDG-1803)
  - ROCS *Ma Kong* (DDG-1805)

Republic of Korea Navy

ROKS Sejong the Great *(DDG-991), a* Sejong the Great-*class destroyer*

- *Sejong the Great*-class destroyer

  - ROKS *Sejong the Great* (DDG-991)
  - ROKS *Yulgok Yi I* (DDG-992)
  - ROKS *Seoae Yu Seong-ryong* (DDG-993)

Royal Navy

- Type 82 destroyer

  - HMS *Bristol* (D23) (now as training ship)

- Type 45 destroyer

  - HMS *Daring* (D32)
  - HMS *Dauntless* (D33)
  - HMS *Diamond* (D34)
  - HMS *Dragon* (D35)
  - HMS *Defender* (D36)
  - HMS *Duncan* (D37)

HMS Daring *(D32), a Type 45 destroyer*

Zumwalt-*class destroyer*

USS Bainbridge *(DDG-96), an* Arleigh Burke-*class destroyer*

United States Navy

- *Zumwalt*-class destroyer
  - USS *Zumwalt* (DDG-1000) (Began sea trials

12/7/15)

- *Arleigh Burke*-class destroyer
  - USS *Arleigh Burke* (DDG-51)
  - USS *Barry* (DDG-52)
  - USS *John Paul Jones* (DDG-53)
  - USS *Curtis Wilbur* (DDG-54)
  - USS *Stout* (DDG-55)
  - USS *John S. McCain* (DDG-56)
  - USS *Mitscher* (DDG-57)
  - USS *Laboon* (DDG-58)
  - USS *Russell* (DDG-59)
  - USS *Paul Hamilton* (DDG-60)
  - USS *Ramage* (DDG-61)
  - USS *Fitzgerald* (DDG-62)
  - USS *Stethem* (DDG-63)
  - USS *Carney* (DDG-64)
  - USS *Benfold* (DDG-65)
  - USS *Gonzalez* (DDG-66)
  - USS *Cole* (DDG-67)
  - USS *The Sullivans* (DDG-68)
  - USS *Milius* (DDG-69)
  - USS *Hopper* (DDG-70)
  - USS *Ross* (DDG-71)
  - USS *Mahan* (DDG-72)
  - USS *Decatur* (DDG-73)
  - USS *McFaul* (DDG-74)
  - USS *Donald Cook* (DDG-75)
  - USS *Higgins* (DDG-76)
  - USS *O'Kane* (DDG-77)
  - USS *Porter* (DDG-78)
  - USS *Oscar Austin* (DDG-79)
  - USS *Roosevelt* (DDG-80)
  - USS *Winston S. Churchill* (DDG-81)
  - USS *Lassen* (DDG-82)
  - USS *Howard* (DDG-83)
  - USS *Bulkeley* (DDG-84)
  - USS *McCampbell* (DDG-85)
  - USS *Shoup* (DDG-86)
  - USS *Mason* (DDG-87)
  - USS *Preble* (DDG-88)
  - USS *Mustin* (DDG-89)

- USS *Chafee* (DDG-90)
- USS *Pinckney* (DDG-91)
- USS *Momsen* (DDG-92)
- USS *Chung-Hoon* (DDG-93)
- USS *Nitze* (DDG-94)
- USS *James E. Williams* (DDG-95)
- USS *Bainbridge* (DDG-96)
- USS *Halsey* (DDG-97)
- USS *Forrest Sherman* (DDG-98)
- USS *Farragut* (DDG-99)
- USS *Kidd* (DDG-100)
- USS *Gridley* (DDG-101)
- USS *Sampson* (DDG-102)
- USS *Truxtun* (DDG-103)
- USS *Sterett* (DDG-104)
- USS *Dewey* (DDG-105)
- USS *Stockdale* (DDG-106)
- USS *Gravely* (DDG-107)
- USS *Wayne E. Meyer* (DDG-108)
- USS *Jason Dunham* (DDG-109)
- USS *William P. Lawrence* (DDG-110)
- USS *Spruance* (DDG-111)
- USS *Michael Murphy* (DDG-112)
- USS *John Finn* (DDG-113)
- USS *Ralph Johnson* (DDG-114)
- USS *Rafael Peralta* (DDG-115)

2.5.2 Former guided missile destroyer classes

France

- *Tourville*-class frigate

Italy

- *Impavido*-class destroyer (decommissioned/retired)
- *Audace*-class destroyer (decommissioned/retired)

Japan

- *Amatsukaze*-class destroyer (decommissioned/retired)
- *Tachikaze*-class destroyer (decommissioned/retired)

Soviet Union

- Kotlin-class destroyer (decommissioned/scrapped)
- Kanin-class destroyer (decommissioned/retired)

United Kingdom

- County-class destroyer (decommissioned/scrapped/sunk)
- Type 42 destroyer (decommissioned/scrapped)

United States

- *Farragut (Coontz)*-class destroyer (decommissioned/scrapped)
- *Charles F. Adams*-class destroyer (all but one sunk for target or scrapped; 1 reserved for future preservation as museum ship)
- *Kidd*-class destroyer (sold to Taiwan as *Kee Lung*-class destroyers)

2.5.3 References

[1] "Indigenously built warship ready for launch". *freepressjournal*. Retrieved 16 April 2015.

[2] "All About the INS Visakhapatnam, Navy's Most Powerful Destroyer". *ndtv*. Retrieved 17 April 2015.

2.6 Huntington Ingalls Industries

Huntington Ingalls Industries (**HII**) is an American Fortune 500 shipbuilding company formed on March 31, 2011 as a spin-off of Northrop Grumman.[1]

It was formerly known as **Northrop Grumman Shipbuilding** (**NGSB**), created on January 28, 2008 by the merger of Northrop Grumman's two shipbuilding sectors, Northrop Grumman Ship Systems and Northrop Grumman Newport News. The company takes its name from the founders of its two main facilities: Collis Potter Huntington (Newport News) and Robert Ingalls (Pascagoula).

Mike Petters is currently the president and CEO of Huntington Ingalls Industries (formerly president of the Newport News shipyard and president of the Northrop Grumman Shipbuilding).[2]

HII is the sole designer, builder, and refueler of nuclear-powered aircraft carriers in the United States. It is one of two nuclear-powered submarine builders (the other being General Dynamics Electric Boat). 70 percent of the current, active US Navy fleet has been built by HII's erstwhile units.

2.6.1 Divisions

- Newport News Shipbuilding, Newport News, Virginia (nuclear aircraft carriers, submarines, refueling and complex overhaul, carrier inactivation)

- Ingalls Shipbuilding, Pascagoula, Mississippi (surface combatants, amphibious warships, Coast Guard large cutters)

2.6.2 Subsidiaries

- AMSEC, Virginia Beach, Virginia (provides maintenance, modernization, logistics, engineering, IT, and training solutions for the U.S. Navy)

- Continental Maritime of San Diego, San Diego, California (Master Ship Repair Contractor for the U.S. Navy and provider of services to Military Sealift Command.)

- Newport News Industrial, Newport News, Virginia (provides fabrication, construction, equipment repair, technical services and products to the energy and petrochemical industries as well as government customers.)

- Stoller Newport News Nuclear (SN3), Broomfield, Colorado (a full-service nuclear operations and environmental services company focused on U.S. Department of Energy (DOE) and U.S. Department of Defense (DoD) clients.)

- Undersea Solutions Group, Panama City Beach, Florida (a leading designer and builder of unmanned underwater vehicles for domestic and international customers.)

- UniversalPegasus International (UPI), Houston, Texas (provides project management, engineering and construction management for the energy industry.)

2.6.3 Facilities

HII operates facilities in several key locations across the US:

- Newport News Shipbuilding, Newport News, Virginia (nuclear aircraft carriers, submarines, overhaul)

- Ingalls Shipbuilding, Pascagoula, Mississippi (surface combatants, amphibs, Coast Guard large cutters)

- Virginia Beach, Virginia (AMSEC, fleet support)

- San Diego, California (Continental Maritime, fleet repair and support)

Former Facilities

- Gulfport, Mississippi (composite R&D, composite components)

- Tallulah, Louisiana (components and subassemblies, closed in 2011)[3]

- Waggaman, Louisiana (closed in 2011)[3]

- Avondale Shipyard, New Orleans, Louisiana (amphibs, auxiliaries, closed in October 2014)[2]

2.6.4 Projects

HII's current order backlog amounts to $22.4 billion.

***Gerald R. Ford*-class aircraft carriers**

HII is to build ten *Gerald R. Ford*-class aircraft carriers for the US Navy. It is scheduled to deliver one carrier every five years starting in 2015.[4]

***America*-class amphibious assault ship**

The US Navy awarded HII a $2.4 billion fixed-price incentive contract for the detail design and construction of the amphibious assault ship *America* (LHA-6), the lead ship of her class. Work will be performed primarily at the company's shipyard in Pascagoula, Miss., and ship delivery is scheduled for 2012.[5]

***San Antonio*-class amphibious transport dock**

In April 2011, the US Navy awarded HII a $1.5 billion contract for the construction of *John P. Murtha* (LPD-26), the tenth of the *San Antonio*-class amphibious transport docks.[6] This was the first Navy contract awarded to HII, though Ingalls Shipbuilding had already built three ships of the class.

Virginia-class attack submarines

The US Navy is building *Virginia*-class submarines as replacements for the Los Angeles-class submarines which are currently being phased out.

HII, under an industrial arrangement with General Dynamics Electric Boat (the only other shipyard capable of building nuclear-powered submarines), solely builds the stern, habitability and machinery spaces, torpedo room, sail and bow, while Electric Boat solely builds the engine room and control room. HII and Electric Boat alternate work on the reactor plant, final assembly, test, outfit and delivery.

Offshore Patrol Cutter

In 2014 The Government Accountability Office denied a contract appeal by Ingalls for the Offshore Patrol Cutter, finding that the USCG's ranking of the shipyard to be marginal was justified.[7]

2.6.5 References

[1] Huntington Ingalls Industries

[2] http://www.sec.gov/Archives/edgar/data/1501585/000150158515000005/hii201410-k.htm

[3] HII 10-K, FY2012, p. 8.

[4] http://www.defenselink.mil/speeches/speech.aspx?speechid=1341

[5] Navy Names New Amphibious Assault Ship

[6] "Ingalls Shipbuilding Awarded U.S. Navy Contract Worth $1.5 Billion to Build Company's 10th San Antonio-Class Amphibious Transport Dock". Huntington Ingalls Industries, Inc. 1 April 2011. Retrieved 2011-04-03.

[7] "GAO denies protest over Coast Guard patrol cutters". *www.washingtontimes.com* (The Associated Press). 1 July 2014. Retrieved 17 August 2015.

2.6.6 External links

- Official website

2.7 Nunn–McCurdy Amendment

The **Nunn–McCurdy Amendment** or **Nunn–McCurdy Provision**, introduced by Senator Sam Nunn and Congressman Dave McCurdy in the United States 1982 Defense Authorization Act and made permanent in 1983, is designed to curtail cost growth in American weapons procurement programs.

It requires notification to the United States Congress if the cost per unit goes more than 25% beyond what was originally estimated, and calls for the termination of programs with total cost growth greater than 50%, unless the Secretary of Defense submits a detailed explanation certifying:

1. the program is essential to national security, that no suitable alternative of lesser cost is available;

2. new estimates of total program costs are reasonable; and

3. management structure is (or has been made) adequate to control costs.

Very rarely is a program actually cancelled under this provision—Congress normally regards the explanations from the Secretary of Defense as acceptable—but it has led to many changes to project management. SBIRS has been affected by the provision in 2002 and again in 2005, and the NPOESS meteorology satellites have been redesigned with lesser capabilities after being affected by the provision.[1][2][3] However, in 2009 and 2011 the US Army's FCS[4] and USMC EFV[5] vehicle programs were both cancelled due to cost overruns.

In 2006, the House of Representatives proposed amending the provision to require a detailed explanation, including information about possible alternatives, at the 15%-cost-growth mark.

2.7.1 References

[1] 109th Congress, 2nd Session, House of Representatives. Hearing before the Committee on Science, *The future of NPOESS: results of the Nunn-McCurdy review of NOAA's weather satellite program*. June 8, 2006. Serial No. 109-53. Government Printing Office, Washington, D.C., 2007

[2] House Report 109-748 - *Summary of activities of the Committee on Science U.S. House of Representatives for the One Hundred Ninth Congress*, p.119. January 4, 2007. Government Printing Office, Washington, D.C., 2007

[3] "NPOESS/Nunn-McCurdy Findings Leave Unanswered Questions on the State of U.S. Weather Forecasting Satellites". *SpaceRef*. 8 June 2006. Retrieved 14 November 2013.

[4] Office of the Assistant Secretary of Defense (Public Affairs) (23 June 2009), "Future Combat System (FCS) Program Transitions to Army Brigade Combat Team Modernization", *United States Department of Defense*, retrieved 13 November 2013

[5] Office of the Assistant Secretary of Defense (Public Affairs) (6 January 2011), "Statement by the Commandant of the Marine Corps Gen. James Amos on Efficiencies", *United States Department of Defense*, retrieved 13 November 2013

2.7.2 External links

- Department of Defense Authorization Act, 1982 at Congress.gov

- S.UP.AMDT.105 at THOMAS

- 2002 Nunn–McCurdy breaches

2.8 Railgun

For railroad artillery, see Railway gun. For other uses, see Rail-gun (disambiguation).
See also: Coilgun
A **railgun** is an electromagnetic projectile launcher based

Naval Surface Warfare Center test firing in January 2008[1]

on similar principles to the homopolar motor. A railgun uses a pair of parallel conductors, or rails, along which a sliding armature is accelerated by the electromagnetic effects of a current that flows down one rail, into the armature and then back along the other rail.[2]

Railguns are being researched as a weapon that would use neither explosives nor propellant, but rather rely on electromagnetic forces to achieve a very high kinetic energy of a projectile. While explosive-powered military guns cannot readily achieve a muzzle velocity of more than about 2

km/s, railguns can readily exceed 3 km/s, and thus far exceed conventionally delivered munitions in range and destructive force. The absence of explosive propellants or warheads to store and handle, as well as the low cost of projectiles compared to conventional weaponry come as additional advantages.[3]

In addition to military applications, NASA has proposed to use a railgun from a high-altitude aircraft to fire a small payload into orbit;[4] however, the extreme g-forces involved would necessarily restrict the usage to only the sturdiest of payloads.

2.8.1 Basics

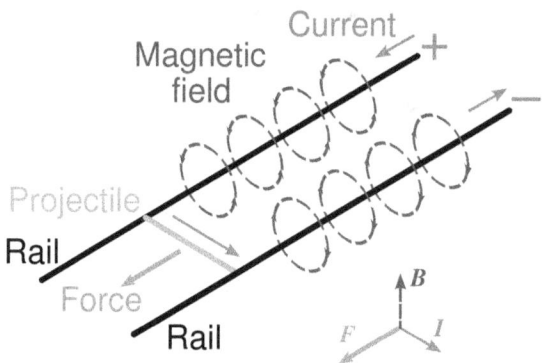

Schematic diagram of a railgun

In its simplest (and most commonly used) form, the railgun differs from a traditional electric motor [5] in that no use is made of additional field windings (or permanent magnets). This basic configuration is formed by a single loop of current and thus requires high currents (e.g. of order one million amperes) to produce sufficient accelerations (and muzzle velocities). A relatively common variant of this configuration is the **augmented railgun** in which the driving current is channelled through additional pairs of parallel conductors, arranged to increase ("augment") the magnetic field experienced by the moving armature.[6] These arrangements reduce the current required for a given acceleration. In electric motor terminology, augmented railguns are usually series-wound configurations.

The armature may be an integral part of the projectile, but it may also be configured to accelerate a separate, electrically isolated or non-conducting projectile. Solid, metallic sliding conductors are often the preferred form of railgun armature but "plasma" or "hybrid" armatures can also be used.[7] A plasma armature is formed by an arc of ionised gas that is used to push a solid, non-conducting payload in a similar manner to the propellant gas pressure in a conventional gun. A hybrid armature uses a pair of "plasma" contacts to interface a metallic armature to the gun rails. Solid armatures may also "transition" into hybrid armatures, typically after a particular velocity threshold is exceeded.

A railgun requires a pulsed, direct current power supply.[8] For potential military applications, railguns are usually of interest because they can achieve much greater muzzle velocities than guns powered by conventional chemical propellants. Increased muzzle velocities can convey the benefits of increased firing ranges while, in terms of target effects, increased terminal velocities can allow the use of kinetic energy rounds as replacements for explosive shells. Therefore, typical military railgun designs aim for muzzle velocities in the range of 2000–3500 m/s with muzzle energies of 5–50 MJ. For comparison, 50MJ is equivalent to the kinetic energy of a school bus weighing 5 metric tons, travelling at 509 km/h (316 mph).[9] For single loop railguns, these mission requirements require launch currents of a few million amperes, so a typical railgun power supply might be designed to deliver a launch current of 5 MA for a few milliseconds. As the magnetic field strengths required for such launches will typically be approximately 10 tesla, most contemporary railgun designs are effectively "air-cored", i.e. they do not use ferromagnetic materials such as iron to enhance the magnetic flux.

It may be noted that railgun velocities generally fall within the range of those achievable by two-stage light-gas guns; however, the latter are generally only considered to be suitable for laboratory use while railguns are judged to offer some potential prospects for development as military weapons. Another light gas gun, the Combustion Light Gas Gun in a 155 mm prototype form was projected to achieve 2500 m/s with a 70 caliber barrel. In some hypervelocity research projects, projectiles are "pre-injected" into railguns, to avoid the need for a standing start, and both two-stage light-gas guns and conventional powder guns have been used for this role. In principle, if railgun power supply technology can be developed to provide compact, reliable and lightweight units, then the total system volume and mass needed to accommodate such a power supply and its primary fuel can become less than the required total volume and mass for a mission equivalent quantity of conventional propellants and explosive ammunition. Such a development would then convey a further military advantage in that the elimination of explosives from any military weapons platform will decrease its vulnerability to enemy fire.

2.8.2   History

In 1918, French inventor Louis Octave Fauchon-Villeplee invented an electric cannon which is an early form of railgun. He filed for a US patent on 1 April 1919, which was issued in July 1922 as patent no. 1,421,435 "Electric Appa-

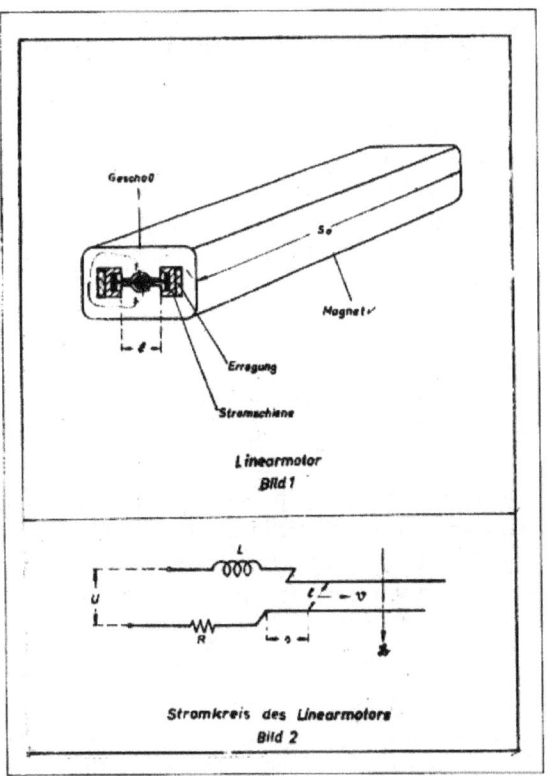

German railgun diagrams

ratus for Propelling Projectiles".[10] In his device, two parallel busbars are connected by the wings of a projectile, and the whole apparatus surrounded by a magnetic field. By passing current through busbars and projectile, a force is induced which propels the projectile along the bus-bars and into flight.[11]

In 1944, during World War II, Joachim Hänsler of Germany's Ordnance Office proposed the first theoretically viable railgun.[12] By late 1944, the theory behind his electric anti-aircraft gun had been worked out sufficiently to allow the Luftwaffe's Flak Command to issue a specification, which demanded a muzzle velocity of 2,000 m/s (6,600 ft/s) and a projectile containing 0.5 kg (1.1 lb) of explosive. The guns were to be mounted in batteries of six firing twelve rounds per minute, and it was to fit existing 12.8 cm FlaK 40 mounts. It was never built. When details were discovered after the war it aroused much interest and a more detailed study was done, culminating with a 1947 report which concluded that it was theoretically feasible, but that each gun would need enough power to illuminate half of Chicago.[11]

During 1950, Sir Mark Oliphant, an Australian physicist and first director of the Research School of Physical Sciences at the new Australian National University, initiated the design and construction of the world's largest (500 megajoule) homopolar generator.[13] This machine was operational from 1962 and was later used to power a large-scale railgun that was used as a scientific experiment.[14]

Late into the first decade of the 2000s, the U.S. Navy tested a railgun that accelerates a 3.2 kg (7 pound) projectile to hypersonic velocities of approximately 2.4 kilometres per second (8,600 km/h), about Mach 7.[15] They gave the project the Latin motto "Velocitas Eradico", Latin for "I, [who am] speed, eradicate" (in the vernacular usage, "Speed Kills").

2.8.3 Design

Theory

A railgun consists of two parallel metal rails (hence the name) connected to an electrical power supply. When a conductive projectile is inserted between the rails (at the end connected to the power supply), it completes the circuit. Electrons flow from the negative terminal of the power supply up the negative rail, across the projectile, and down the positive rail, back to the power supply.[16]

This current makes the railgun behave as an electromagnet, creating a magnetic field inside the loop formed by the length of the rails up to the position of the armature. In accordance with the right-hand rule, the magnetic field circulates around each conductor. Since the current is in the opposite direction along each rail, the net magnetic field between the rails (**B**) is directed at right angles to the plane formed by the central axes of the rails and the armature. In combination with the current (**I**) in the armature, this produces a Lorentz force which accelerates the projectile along the rails, away from the power supply. There are also Lorentz forces acting on the rails and attempting to push them apart, but since the rails are mounted firmly, they cannot move.

By definition, if a current of one ampere flows in a pair of infinitely long parallel conductors that are separated by a distance of one metre, then the magnitude of the force on each metre of those conductors will be exactly 0.2 micronewtons. Furthermore, in general, the force will be proportional to the square of the magnitude of the current and inversely proportional to the distance between the conductors. It also follows that, for railguns with projectile masses of a few kg and barrel lengths of a few m, very large currents will be required to accelerate projectiles to velocities of the order of 1000 m/s.

A very large power supply, providing on the order of one million amperes of current, will create a tremendous force on the projectile, accelerating it to a speed of many kilometres per second (km/s). 20 km/s has been achieved with small projectiles explosively injected into the railgun. Although these speeds are possible, the heat generated from

the propulsion of the object is enough to erode the rails rapidly. Under high-use conditions, current railguns would require frequent replacement of the rails, or to use a heat-resistant material that would be conductive enough to produce the same effect. At this time it is generally acknowledged that it will take major breakthroughs in material science and related disciplines to produce high-powered railguns capable of firing more than a few shots from a single set of rails. The barrel must withstand these conditions for up to several rounds per minute for thousands of shots without failure or significant degradation. These parameters are well beyond the state of the art in materials science.[17]

Mathematical formula

The magnitude of the force vector can be determined from a form of the Biot–Savart law and a result of the Lorentz force. It can be derived mathematically in terms of the permeability constant (μ_0), the radius of the rails (which are assumed to be circular in cross section) (r), the distance between the centrepoints of the rails (d) and the current in amps through the system (I) as follows:

It can be shown from the Biot-Savart law that at one end of a semi-infinite current-carrying wire, the magnetic field at a given perpendicular distance (s) from the end of the wire is given by:[18]

$$\mathbf{B}(s) = \frac{\mu_0 I}{4\pi s} \hat{\phi}$$

Note this is if the wire runs from the location of the armature e.g. from x = 0 back to $x = -\infty$ and s is measured relative to the axis of the wire.

So, if the armature connects the ends of two such semi-infinite wires separated by a distance, d , a fairly good approximation assuming the length of the wires is much larger than d , the total field from both wires at any point on the armature, or any point in the plane between the two wires is:

$$B(s) = \frac{\mu_0 I}{4\pi} \left(\frac{1}{s} + \frac{1}{d-s} \right) \hat{z}$$

Where s is the perpendicular distance from the point on the armature to the axis of one of the wires.

Note that $\hat{\phi}$ between the rails is $\hat{z}$ assuming the rails are lying in the xy plane and run from x = 0 back to $x = -\infty$ as suggested above.

To obtain an approximate expression for the force on the railgun armature, we start by again assuming that the railgun rails can be modeled as a pair of semi-infinite conductors.

This allows us to use the above expression for the magnetic field on the armature in the Lorentz Force Law,

$$\mathbf{F} = I \int \mathrm{d}\boldsymbol{\ell} \times \mathbf{B}$$

Inserting the expression for the magnetic field into the Lorentz force law and setting the bounds of integration to r and $d - r$, assuming the armature is a bar between and perpendicular to the rails making $\mathrm{d}\boldsymbol{\ell}$ point in the $\hat{y}$, we find the force on the armature is

$$\mathbf{F} = I \int_{r}^{d-r} \mathrm{d}\boldsymbol{\ell} \times \frac{\mu_0 I}{4\pi} \left(\frac{1}{s} + \frac{1}{d-s} \right) \hat{z} = \frac{\mu_0 I^2}{2\pi} \ln \left(\frac{d-r}{r} \right) \hat{x}$$

The formula is based on the assumption that the distance (l) between the point where the force (F) is measured and the beginning of the rails is greater than the separation of the rails (d) by a factor of about 3 or 4 ($l > 3d$). Some other simplifying assumptions have also been made; to describe the force more accurately, the geometry of the rails and the projectile must be considered.

With most practical railgun geometries, it is not easy to produce an electromagnetic expression for the railgun force that is both simple and reasonably accurate. Instead, most practical railgun analyses actually used a lumped circuit model to describe the relationship between the driving current and the railgun force. In these models the voltage across the railgun breech is given by:

$$V = IR + \frac{\mathrm{d}(LI)}{\mathrm{d}t}$$

Then the barrel resistance and inductance are assumed to vary linearly with the projectile position, so that

$$R = R'x$$

$$L = L'x$$

from which

$$V = I(R'x + L'v) + L'x\frac{\mathrm{d}I}{\mathrm{d}t}$$

If the driving current is held constant, there is a power flow equal to $I^2 L'v$ which represents the electromagnetic work done. In this simple model, exactly half of this is assumed to be needed to establish the magnetic field along the barrel, i.e. as the length of the current loop increases. The other half represents the power flow into the kinetic energy of the projectile. Since power can be expressed as force times speed, this gives the standard result that the force on the railgun armature is given by:

$$F = \frac{L'I^2}{2}$$

This simple equation shows that high accelerations will require very high currents. For an ideal square bore railgun, the value of L' would be about 0.6 microHenries per metre (μ.H/m) but most practical railgun barrels exhibit lower values of L' than this.

Since the lumped circuit model describes the railgun force in terms of fairly normal circuit equations, it becomes possible to specify a simple time domain model of a railgun. Ignoring friction and air drag, the projectile acceleration is given by:

$$\frac{\mathrm{d}v}{\mathrm{d}t} = \frac{L'I^2}{2m}$$

where m is the projectile mass. The motion along the barrel is given by:

$$\frac{\mathrm{d}x}{\mathrm{d}t} = v$$

and the above voltage and current terms can be placed into appropriate circuit equations to determine the time variation of current and voltage.

It can also be noted that the textbook formula for the high frequency inductance per unit length of a pair of parallel round wires, of radius r and axial separation d is:

$$L' = \frac{\mu_0}{\pi} \ln \left(\frac{d-r}{r} \right)$$

so the lumped parameter model also predicts the force for this case as:

$$F = \frac{L'I^2}{2} = \frac{\mu_0 I^2}{2\pi} \ln \left(\frac{d-r}{r} \right) .$$

With practical railgun geometries, much more accurate two or three dimensional models of the rail and armature current distributions (and the associated forces) can be computed, e.g. by using finite element methods to solve formulations based on either the scalar magnetic potential or the magnetic vector potential.

Design considerations

The power supply must be able to deliver large currents, sustained and controlled over a useful amount of time. The most important gauge of power supply effectiveness is the energy it can deliver. As of December 2010, the greatest known energy used to propel a projectile from a railgun was 33 megajoules.[19] The most common forms of power supplies used in railguns are capacitors and compulsators which are slowly charged from other continuous energy sources.

The rails need to withstand enormous repulsive forces during shooting, and these forces will tend to push them apart and away from the projectile. As rail/projectile clearances increase, arcing develops, which causes rapid vaporization and extensive damage to the rail surfaces and the insulator surfaces. This limited some early research railguns to one shot per service interval.

The inductance and resistance of the rails and power supply limit the efficiency of a railgun design. Currently different rail shapes and railgun configurations are being tested, most notably by the United States Navy, the Institute for Advanced Technology at the University of Texas at Austin, and BAE Systems.

Materials used

The rails and projectiles must be built from strong conductive materials; the rails need to survive the violence of an accelerating projectile, and heating due to the large currents and friction involved. Some erroneous work has suggested that the recoil force in railguns can be redirected or eliminated; careful theoretical and experimental analysis reveals that the recoil force acts on the breech closure just as in a chemical firearm.[20][21][22][23] The rails also repel themselves via a sideways force caused by the rails being pushed by the magnetic field, just as the projectile is. The rails need to survive this without bending and must be very securely mounted. Currently published material suggests that major advances in material science must be made before rails can be developed that allow railguns to fire more than a few full-power shots before replacement of the rails is required.

Heat dissipation

In current designs massive amounts of heat are created by the electricity flowing through the rails, as well as by the friction of the projectile leaving the device. The heat created by this friction itself can cause thermal expansion of the rails and projectile, further increasing the frictional heat. This causes three main problems: melting of equipment, decreased safety of personnel, and detection by enemy forces due to increased infrared signature. As briefly discussed above, the stresses involved in firing this sort of device require an extremely heat-resistant material. Otherwise the rails, barrel, and all equipment attached would melt or be irreparably damaged.

In practice the rails are, with most designs of railgun, subject to erosion due to each launch; in addition, projectiles can be subject to some degree of ablation, and this can limit railgun life, in some cases severely.[24]

2.8.4   Applications

Railguns have a number of potential practical applications, primarily for the military. However, there are other theoretical applications currently being researched.

Launch or launch assist of spacecraft

Main article: Mass driver
See also: Space gun

Electrodynamic assistance to launch rockets has been studied.[25] Space applications of this technology would likely involve specially formed electromagnetic coils and superconducting magnets.[26] Composite materials would likely be used for this application.[27]

For space launches from Earth, relatively short acceleration distances (less than a few km) would require very strong acceleration forces, higher than humans can tolerate. Other designs include a longer helical (spiral) track, or a large ring design whereby a space vehicle would circle the ring numerous times, gradually gaining speed, before being released into a launch corridor leading skyward.

In 2003, Ian McNab outlined a plan to turn this idea into a realized technology.[28] Because of strong acceleration, this system would launch only sturdy materials, such as food, water, and – most importantly – fuel. Under ideal circumstances (equator, mountain, heading east) the system would cost $528/kg,[28] compared with $5,000/kg on the conventional rocket.[29] The McNab's railgun could make approximately 2000 launches per year, for a total of maximum 500 tons launched per year. Because the launch track would be 1.6 km long, power will be supplied by a distributed network of 100 rotating machines (compulsator) spread along the track. Each machine would have a 3.3-ton carbon fibre rotor spinning at high speeds. A machine can recharge in a matter of hours using 10 MW power. This machine could be supplied by a dedicated generator. The total launch package would weigh almost 1.4 tons. Payload per launch in these conditions is over 400 kg.[28] There would be a peak operating magnetic field of 5 T—half of this coming from the rails, and the other half from augmenting magnets. This halves the required current through the rails, which reduces the power fourfold.

Weaponry

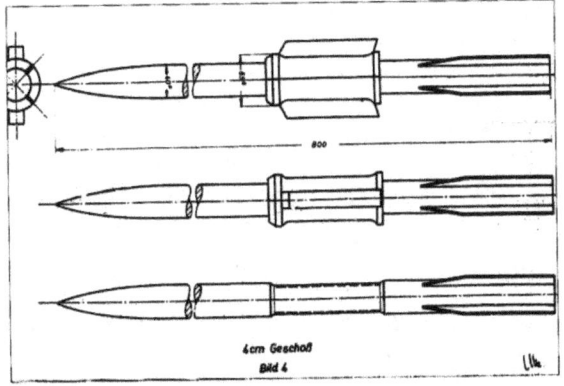

Drawings of electric gun projectiles

Railguns are being researched as weapons with projectiles that do not contain explosives or propellants, but are given extremely high velocities: 2,500 m/s (8,200 ft/s) (approximately Mach 7 at sea level) or more. For comparison, the M16 rifle has a muzzle speed of 930 m/s (3,050 ft/s), and the 16"/50 caliber Mark 7 gun that armed World War II American battleships has a muzzle speed of 760 m/s (2,490 ft/s)), which because of its much greater mass generated a muzzle energy of 360 MJ and a downrange kinetic impact of energy of over 160 MJ. By firing smaller projectiles at extremely high velocities, railguns can yield kinetic energy impacts equal or superior to the destructive energy of 5" Naval guns, but with much greater range. This decreases ammunition size and weight, allowing more ammunition to be carried and eliminating the hazards of carrying explosives or propellants in a tank or naval weapons platform. Also, by firing at greater velocities, railguns have greater range, less time to target, and at shorter ranges less wind drift, bypassing the physical limitations of conventional firearms: *"the limits of gas expansion prohibit launching an unassisted projectile to velocities greater than about 1.5 km/s and ranges of more than 50 miles [80 km] from a practical conventional gun system."*[30] Current railgun technologies necessitate a long and heavy barrel, but a railgun's ballistics far outperform conventional cannons of equal barrel lengths. Railguns can also deliver area of effect damage by detonating a bursting charge in the projectile which unleashes a swarm of smaller projectiles over a large area.[31][32]

Assuming that the many technical challenges facing fieldable railguns are overcome, including tough ones like railgun projectile guidance and rail endurance, the increased launch velocities of railguns will provide advantages over more conventional guns for a variety of offensive and defensive scenarios. Railguns have the potential to be used against both surface and airborne targets.

Many critics of weaponized railgun systems claim operating them with a suitable exit velocity and rate of fire would consume too much power, though this would likely not be a problem for nuclear-powered systems such as on large warships or submarines.

The first weaponized railgun planned for production, the General Atomics Blitzer system, began full system testing in September 2010. The weapon launches a streamlined discarding sabot round designed by Boeing's Phantom Works at 1,600 m/s (5,200 ft/s) (approximately Mach 5) with accelerations exceeding 60,000 g_n.[33] During one of the tests, the projectile was able to travel an additional 7 kilometres (4.3 mi) downrange after penetrating a $\frac{1}{8}$ inch (3.2 mm) thick steel plate. The company hopes to have an integrated demo of the system by 2016 followed by production by 2019, pending funding. Thus far, the project is self-funded.[34]

In October 2013, General Atomics unveiled a land based version of the Blitzer railgun. A company official claimed the gun could be ready for production in "two to three years".[35]

Railguns are being examined for use as anti-aircraft weapons to intercept air threats, particularly anti-ship cruise missiles, in addition to land bombardment. A supersonic sea-skimming anti-ship missile can appear over the horizon 20 miles from a warship, leaving a very short reaction time for a ship to intercept it. Even if conventional defense systems react fast enough, they are expensive and only a limited number of large interceptors can be carried. A railgun projectile can reach several times the speed of sound faster than a missile, so it can reach out to the horizon and hit a cruise missile much faster and further away from the ship. Projectiles are also cheaper and smaller, allowing for many more to be carried. The speed, cost, and numerical advantages of railgun systems may allow them to replace several different systems in the current layered defense approach.[36] A railgun projectile without the ability to change course can hit fast-moving missiles at a maximum range of 30 nmi (35 mi; 56 km).[37] As is the case with the Phalanx CIWS, unguided railgun rounds will require multiple/many shots to bring down maneuvering supersonic anti-ship missiles, with the odds of hitting the missile improving dramatically the closer it gets. The Navy plans for railguns to be able to intercept endo-atmospheric ballistic missiles, stealthy air threats, supersonic missiles, and swarming surface threats; a prototype system for supporting interception tasks is to be ready by 2018, and operational by 2025. This timeframe suggests the weapons are planned to be installed on the Navy's next-generation surface combatants, expected to start construction by 2028.[38]

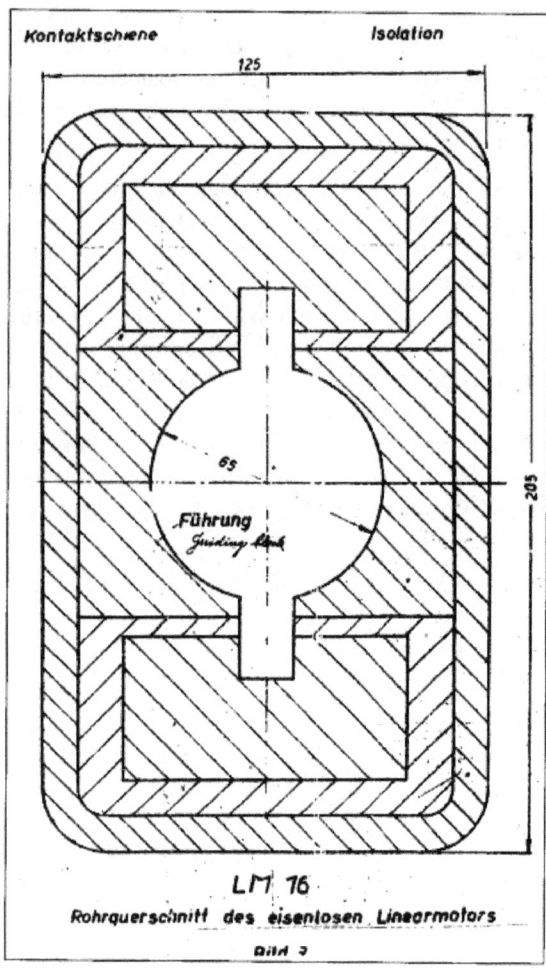

Diagram showing the cross-section of a linear motor cannon

achieve a speed of 4,500 m/s (14,800 ft/s). The aim was to achieve projectile speed of 7,000 m/s (23,000 ft/s). At the time, it was considered a military secret.

China is now one of the major players in electromagnetic launchers; in 2012 it hosted the 16th International Symposium on Electromagnetic Launch Technology (EML 2012) at Beijing.[39] Satellite imagery in late 2010 suggested that tests were being conducted at an armor and artillery range near Baotou, in the Inner Mongolia Autonomous Region.[40]

U.S. military tests   The United States military is funding railgun experiments. At the University of Texas at Austin Center for Electromechanics, military railguns capable of delivering tungsten armor-piercing bullets with kinetic energies of nine megajoules have been developed.[41] 9 MJ is enough energy to deliver 2 kg (4.4 lb) of projectile at 3 km/s (1.9 mi/s)—at that velocity a rod of tungsten or another dense metal could easily penetrate a tank, and potentially pass through it.

U.S. Navy tests   The United States Naval Surface Warfare Center Dahlgren Division demonstrated an 8 MJ railgun firing 3.2 kg (7.1 lb) projectiles in October 2006 as a prototype of a 64 MJ weapon to be deployed aboard Navy warships. The main problem the U.S. Navy has had with implementing a railgun cannon system is that the guns wear out due to the immense pressures, stresses and heat that are generated by the millions of amperes of current necessary to fire projectiles with megajoules of energy. Such weapons, while not nearly as powerful as a cruise missile like a BGM-109 Tomahawk cruise missile that will deliver 3000 MJ of destructive energy to a target, will theoretically allow the Navy to deliver more granular firepower at a fraction of the cost of a missile, and will be much harder to shoot down versus future defensive systems. For context another relevant comparison is the Rheinmetall 120mm gun used on main battle tanks will generate 9 MJ of muzzle energy. An MK 8 round fired from the 16" guns of an Iowa Class battleship at 2500 fps (762 m/s) has 360 MJ of kinetic energy at the muzzle.

Since then, BAE Systems has delivered a 32 MJ prototype (muzzle energy) to the U.S. Navy.[42] The same amount of energy is released by the detonation of 4.8 kg (11 lb) of C4.

On January 31, 2008 the U.S. Navy tested a railgun that fired a projectile at 10.64 MJ with a muzzle velocity of 2,520 m/s (8,270 ft/s).[43] The power was provided by a new 9-megajoule prototype capacitor bank using solid-state switches and high-energy-density capacitors delivered in 2007 and an older 32-MJ pulse power system from the US Army's Green Farm Electric Gun Research and Develop-

Tests   Full-scale models have been built and fired, including a 90 mm (3.5 in) bore, 9 MJ kinetic energy gun developed by the US DARPA. Rail and insulator wear problems still need to be solved before railguns can start to replace conventional weapons. Probably the oldest consistently successful system was built by the UK's Defence Research Agency at Dundrennan Range in Kirkcudbright, Scotland. This system was established in 1993 and has been operated for over 10 years. Using its associated flight range for internal, intermediate, external and terminal ballistics, it achieved several mass and velocity records.

The Yugoslavian Military Technology Institute developed, within a project named EDO-0, a railgun with 7 kJ kinetic energy, in 1985. In 1987 a successor was created, project EDO-1, that used projectile with a mass of 0.7 kg (1.5 lb) and achieved speeds of 3,000 m/s (9,800 ft/s), and with a mass of 1.1 kg (2.4 lb) reached speeds of 2,400 m/s (7,900 ft/s). It used a track length of 0.7 m (2.3 ft). According to those working on it, with other modifications it was able to

ment Facility developed in the late 1980s that was previously refurbished by General Atomics Electromagnetic Systems (EMS) Division.[44] It is expected to be ready between 2020 to 2025.[45]

A test of a railgun took place on December 10, 2010, by the US Navy at the Naval Surface Warfare Center Dahlgren Division.[46] During the test, the Office of Naval Research set a world record by conducting a 33 MJ shot from the railgun, which was built by BAE Systems.[19][47]

A test took place in February, 2012, at the Naval Surface Warfare Center Dahlgren Division. While similar in energy to the aforementioned test, the railgun used is considerably more compact, with a more conventional looking barrel. A General Atomics-built prototype was delivered for testing in October 2012.[48]

The U.S. Navy plans to integrate a railgun that has a range of over 160 km (100 mi) onto a ship by 2016.[49] This weapon, while having a form factor more typical of a naval gun will utilize components largely in common with those developed and demonstrated at Dahlgren.[50] The hypervelocity rounds weigh 10 kg (23 lb), are 18 in (460 mm), and are fired at Mach 7.[51]

A future goal is to develop projectiles that are self-guided - a necessary requirement to hit distant targets or intercepting missiles.[52] When the guided rounds are developed, the Navy is projecting each round to cost about $25,000,[53] though it must be noted that developing guided projectiles for guns has a history of doubling or tripling initial cost estimates. Some HPV projectiles developed by the Navy have command guidance, but the accuracy of the command guidance is not known, nor even if it can survive a full power shot.

Currently the only US Navy ships that can produce enough electrical power to get desired performance are the Zumwalt-class destroyers; they can generate 78 megawatts of power, more than is necessary to power a railgun. Engineers are working to derive technologies developed for the DDG-1000 series ships into a battery system so other warships can operate a railgun.[54] Most current destroyers can spare only nine megawatts of additional electricity, while it would require 25 megawatts to propel a projectile to the desired maximum range [55] (i.e. to launch 32MJ projectiles at a rate of 10 shots per minute). Even if current ships, such as the Arleigh Burke-class destroyer, can be upgraded with enough electrical power to operate a railgun, the space taken up on the ships by the integration of an additional weapon system may force the removal of existing weapon systems to make room available.[56] The first shipboard tests will be from a railgun installed on an Expeditionary Fast Transport. Though ships of that class are non-combatants, they were chosen for their available cargo and topside space and schedule flexibility. They will not be permanently in-

stalled on the EFT, and the Navy has yet to decide which ship classes will receive a fully operational railgun.[57] Single shot tests will be held in 2016, followed by an autoloader in 2018.[58] According to the Navy "current research is focused on a rep-rate capability of multiple rounds per minute which entails development of a tactical prototype gun barrel and pulsed power systems incorporating advanced cooling techniques. Components are designed to transition directly into prototype systems now being conceptualized."[50] So as of March 2014 multiple rep railgun were in the conceptual stage and have a ways to go before reaching the prototype stage.

Though the 23 lb projectiles have no explosives, their Mach 7 velocity gives them 32 megajoules of energy, but impact kinetic energy downrange will typically be 50 percent or less of the muzzle energy. The Navy is looking into other uses for railguns, besides land bombardment, such as air defense; with the right targeting systems, projectiles could intercept aircraft, cruise missiles, and even ballistic missiles. The Navy is also developing directed energy weapons for air defense use, but it will be years before they will be effective.

The railgun will be part of a Navy fleet that envisions future offensive and defensive capabilities being provided in layers: lasers to provide close range defense, railguns to provide medium range attack and defense, and cruise missiles to provide long-range attack; though railguns will cover targets up to 100 miles away that previously needed a missile.[59]

The Navy may eventually enhance railgun technology to enable it to fire at a range of 200 nmi (230 mi; 370 km) and impact with 64 megajoules of energy. One shot would require 6 million amps of current, so it will take a long time to develop capacitors that can generate enough energy and strong enough gun materials.[40]

Outstanding Issues in Fielding Railgun Weapons   Major technological and operational hurdles must be overcome before railguns can be deployed:

1) **Railgun durability:** To date railgun demonstrations, while impressive, have not demonstrated an ability to fire multiple full power shots from the same set of rails. The Navy has claimed hundreds of shots from the same set of rails. In a March 2014 statement to the Intelligence, Emerging Threats and Capabilities Subcommittee of the House Armed Services Committee, Chief of Naval Research Admiral Matthew Klunder stated, "Barrel life has increased from tens of shots to over 400, with a program path to achieve 1000 shots."[50] However, the Office of Naval Research (ONR) will not confirm that the 400 shots are full-power shots. Further there is nothing published to indicate there are any high megajoule class railguns with the capability of firing hundreds of full-power shots while staying

within the strict operational parameters necessary to fire railgun shots accurately and safely. As noted in an article by Globalsecurity.org:[60] railguns should be able to fire 6 rounds per minute with a rail life of about 3000 rounds. Given launch acceleration of up to 60,000 g's, massive pressures and mega amps of current, railgun rails are quickly destroyed and getting to the endurance to fire hundreds of full-power rounds, to say nothing of thousands of rounds, could require breakthroughs in materials science that cannot be scheduled and could be decades in coming.

Until the capability of firing at least hundreds of rounds of full power shots from the same set of rails is demonstrated, railguns as fieldable weapons remain an interesting idea with a lot of potential.

2) **Railgun Projectile Guidance:** While the Navy has made reference to having successfully integrated "command guidance" into railgun projectiles there is no published documentation of having successfully tested such a capability. Command guidance, as opposed to self-guidance, involves direct control of the railgun projectile by the launching authority using such technologies as radio or wire, etc. With the kind of speeds obtained by railgun projectiles and the ranges possible, command guidance will be of limited use. Further, no details have been given as to effectiveness of the command guidance or if it will actually work with full power railgun shots. Until the Navy releases more information, command guidance effectiveness is an unknown.

3) **Self-guided Railgun Projectiles:** A future capability critical to fielding a real railgun weapon is developing a robust guidance package that will allow the railgun to fire at distant targets or to hit incoming missiles. Developing such a package is a real challenge. The Navy's RFP Navy SBIR 2012.1 - Topic N121-102 [61] for developing such a package gives a good overview of just how challenging railgun projectile guidance is:

"The package must fit within the mass (< 2 kg), diameter (< 40 mm outer diameter), and volume (200 cm^3) constraints of the projectile and do so without altering the center of gravity. It should also be able to survive accelerations of at least 20,000 g (threshold) / 40,000 g (objective) in all axes, high electromagnetic fields (E > 5,000 V/m, B > 2 T), and surface temperatures of > 800 deg C. The package should be able to operate in the presence of any plasma that may form in the bore or at the muzzle exit and must also be radiation hardened due to exo-atmospheric flight. Total power consumption must be less than 8 watts (threshold) / 5 watts (objective) and the battery life must be at least 5 minutes (from initial launch) to enable operation during the entire engagement. In order to be affordable, the production cost per projectile must be as low as possible, with a goal of less than $1,000 per unit."

While the specifications mention 40,000 g's, actual railgun launches can reach 60,000 g's so this contract is to develop a base capability that would be appropriate for launches of less than full power pulling less than 40,000 g's.

On June 22, 2015, General Atomics' Electromagnetic Systems announced that projectiles with on-board electronics survived the whole railgun launch environment and performed their intended functions in four consecutive tests on June 9 and 10 June at the U.S. Army's Dugway Proving Ground in Utah. The on-board electronics successfully measured in-bore accelerations and projectile dynamics, for several kilometers downrange, with the integral data link continuing to operate after the projectiles impacted the desert floor, which is essential for precision guidance.[62]

Trigger for inertial confinement fusion

Railguns may also be miniaturized for inertial confinement nuclear fusion.

- Fusion is triggered by very high temperature and pressure at the core.

  - Current technology calls for multiple lasers, usually over 100, to concurrently strike a fuel pellet, creating a symmetrical compressive pressure.

  - Railguns may be able to trigger fusion by firing energetic plasma from multiple directions. The process developed involves four key steps.[63]

    - Plasma is pumped into a chamber.
    - When the pressure is great enough, a diaphragm will rupture, sending gas down the rail.
    - Shortly afterwards, a sufficient voltage is applied to the rails, creating a conduction path of ionized gas.
    - This plasma is accelerated down the rail, eventually being ejected at a large velocity.

- The rails and dimensions are on the order of centimetres.

2.8.5 See also

- Coilgun

- Homopolar generator

- Kinetic energy penetrator

- Light-gas gun

- List of caseless firearms

- List of electromagnetic projectile devices in fiction

- Mass driver

- Non-rocket spacelaunch

- Project Babylon

- Ram accelerator

- Space gun

- USNS Trenton (T-EPF-5), first ship to mount a railgun.[64]

2.8.6   References

[1] Fletcher, Seth (2013-06-05). "Navy Tests 32-Megajoule Railgun I". *Popular Science*. Retrieved 2013-06-16.

[2] Rashleigh, C. S. & Marshall, R. A. (April 1978). "Electromagnetic Acceleration of Macroparticles to High Velocities". *J. Appl. Phys.* **49** (4): 2540. doi:10.1063/1.325107.

[3] "Rail Strike". *The Economist*. 2015-05-09. Retrieved 2016-01-31.

[4] Atkinson, Nancy (2010-09-14). "NASA Considering Rail Gun Launch System to the Stars". *Universe Today*.

[5] Hindmarsh, John (1977). *Electrical Machines and their Applications*. Oxford: Pergamon Press. p. 20. ISBN 0-08-021165-8.

[6] Fiske, D.; Ciesar, J.A.; Wehrli, H.A.; Riemersma, H.; et al. (January 1991). "The HART 1 Augmented Electric Gun Facility". *IEEE Transactions on Magnetics* **27** (1): 176–180. doi:10.1109/20.101019. ISSN 0018-9464.

[7] Batteh, Jad. H. (January 1991). "Review of Armature Research". *IEEE Transactions on Magnetics* (IEEE Magnetics Society) **27** (1): 224–227. doi:10.1109/20.101030.

[8] Gully, John (January 1991). "Power Supply Technology for Electric Guns". *IEEE Transactions on Magnetics* (IEEE Magnetics Society) **27** (1): 329–334. doi:10.1109/20.101051.

[9] "50 megajoules kinetic energy". *Wolfram Alpha*. 2014-04-28.

[10] Fauchon-Villeplee, André Louis Octave (1922). "US Patent 1,421,435 "Electric Apparatus for Propelling Projectiles"".

[11] Hogg, Ian V. (1969). *The Guns: 1939/45*. London: Macdonald. ISBN 9780019067102. OCLC 778837078.

[12] http://faculty.kfupm.edu.sa/EE/husainm/EE%20340/ Research%20Projects/Students%20Reports/6-%20Railguns-%20Al-Khaldi.pdf

[13] Ophel, Trevor & Jenkin, John (1996). "Chapter 2:The Big Machine" (PDF). *Fire in the Belly: The first fifty years of the pioneer School at the ANU*. Australian National University. ISBN 9780858000483. OCLC 38406540. Retrieved 2014-04-10.

[14] Barber, J. P. (March 1972). *The Acceleration of Macroparticles and a Hypervelocity Electromagnetic Accelerator* (Ph.D thesis). Australian National University. OCLC 220999609.

[15] Borrell, Brendan (2008-02-06). "Electromagnetic Railgun Blasts Off". *MIT Technology Review*.

[16] Harris, William (11 October 2005). "How Rail Guns Work". *HowStuffWorks*. Retrieved 2011-03-25.

[17] "Electromagnetic Rail Gun (EMRG)". *globalsecurity.org*.

[18] Smolinski, Jason. "Magnetism". Retrieved 2014-09-04.

[19] Ackerman, Spencer (2010-12-10). "Video: Navy's Mach 8 Railgun Obliterates Record". *Wired*.

[20] Weldon, Wm. F.; Driga, M. D. & Woodson, H. H. (November 1986). "Recoil in electromagnetic railguns". *IEEE Transactions on Magnetics* **22** (6): 1808–1811. doi:10.1109/TMAG.1986.1064733. ISSN 0018-9464.

[21] Cavalleri, G.; Tonni, E. & Spavieri, G. (May 2001). "Reply to "Electrodynamic force law controversy"". *Physical Review E* **63** (5): 058602. Bibcode:2001PhRvE..63e8602C. doi:10.1103/PhysRevE.63.058602.

[22] Kathe, Eric L. (November 2000). *Recoil Considerations for Railguns: Technical Report ARCCB-TR-00016* (pdf). U.S. Army ARDEC Benet Laboratories.

[23] Putnam, Michael J. (December 2009). *An Experimental Study of Electromagnetic Lorentz Force and Rail Recoil* (M.Sc. thesis). Naval Postgraduate School.

[24] Barros, Sam (2010-11-11). "PowerLabs Rail Gun!". Powerlabs.org (Blog). Retrieved 2014-04-10.

[25] Uranga, Alejandra; Kirk, Daniel R.; Gutierrez, Hector; Meinke, Rainer B.; et al. (2005). *Rocket Performance Analysis Using Electrodynamic Launch Assist* (PDF). Proceedings of the 43rd AIAA Aerospace Sciences Meeting and Exhibit (10–13 January 2005). Reno, Nevada.

[26] Advanced Magnet Lab, Inc. (2008) "Space and Defense" *magnetlab.com* Archived October 14, 2008, at the Wayback Machine.

[27] Advanced Magnet Lab, Inc. (2008) "Direct Double-Helix" *magnetlab.com* Archived February 13, 2011, at the Wayback Machine.

[28] McNab, I.R. (January 2003). "Launch to space with an electromagnetic railgun" (PDF). *IEEE Transactions on Magnetics* **35** (1): 295–304. doi:10.1109/TMAG.2002.805923. ISSN 0018-9464.

[29] Proton is estimated at $5000/kg as of 2015.

[30] Adams, David Allan (February 2003). "Naval Rail Guns Are Revolutionary" (PDF). *U.S. Naval Institute Proceedings* **129** (2): 34. Archived from the original (PDF) on 2007-07-08.

[31] "Railguns". *http://navy-matters.blogspot.com/*. Navy Matters. Retrieved 11 February 2015. External link in |website= (help)

[32] Fredenburg, Michael. "Railguns: The Next Big Pentagon Boondoggle?". *www.nationalreview.com*. National Review.

[33] Fallon, Jonathon (2012-04-25). "General Atomics' Railgun Travels 4 Miles, Even After Blasting Through a Steel Plate [Video]". CubicleBot. Retrieved 2012-04-25.

[34] "Blitzer Railgun". General Atomics. 2012-04-25. Retrieved 2012-04-25.

[35] Fisher Jr, Richard D. (2013-10-22). "AUSA 2013: General Atomics unveils Blitzer land-based railgun". Jane's. Archived from the original on 2014-03-29. Retrieved 2014-12-22.

[36] Page, Lewis (2010-12-25). "'Blitzer' railgun already 'tactically relevant', boasts maker". *The Register*.

[37] Freedberg Jr., Sydney J. (2014-11-21). "47 Seconds From Hell: A Challenge To Navy Doctrine". *Breaking Defense*.

[38] LaGrone, Sam (2015-01-05). "Navy Wants Rail Guns to Fight Ballistic and Supersonic Missiles Says RFI". *USNI News*.

[39] LIST OF PAPERS, 16th International Symposium on Electromagnetic Launch Technology (EML 2012) Beijing, China, ISBN 978-1-4673-0306-4, http://toc.proceedings.com/16118webtoc.pdf

[40] Five Futuristic Weapons That Could Change Warfare - Nationalinterest.org, 1 November 2014

[41] "EM Systems". University of Texas. Archived from the original on 2007-10-10.

[42] Sofge, Erik (2007-11-14). "World's Most Powerful Rail Gun Delivered to Navy". Popular Mechanics. Retrieved 2007-11-15.

[43] "U.S. Navy Demonstrates World's Most Powerful EMRG at 10 MJ". *United States Navy*. 2008-02-01.

[44] "General Atomics Team Powers Navy Rail Gun to New World Record, accessed 14 Oct 2009"

[45] "The Navy shows off its insane magnetic railgun of the future". Dvice.com. 2008-02-02. Retrieved 2014-04-10.

[46] Fein, Geoff. "Navy Sets New World Record with Electromagnetic Railgun Demonstration". *http://www.navy.mil/*. United States Navy. Retrieved 13 February 2015. External link in |website= (help)

[47] LaGrone, Sam (2010-12-15). "Electromagnetic railgun sets new world record". *Jane's*. Archived from the original on 2010-12-17. Retrieved 2014-12-22.

[48] "Navy Evaluating Second Electromagnetic Railgun Innovative Naval Prototype". *Office of Naval Research*. 2012-10-09. Retrieved 2012-10-20.

[49] Osborn, Kris (2014-01-10). "Future Destroyers Likely to Fire Lasers, Rail Guns". *Military.com*.

[50] Klunder, Matthew. "Statement of Read Admiral Matthew L. Klunder, United States Navy Chief of Naval Research Before the Intelligence, Emerging Threats and Capabilities Subcommittee of the House Armed Services Committee on the Fiscal Year 2015 Budget Request" (PDF). *www.acq.osd.mil*. House Armed Services Committee. Retrieved 13 February 2015.

[51] McDuffee, Allen (2014-04-09). "Navy's New Railgun Can Hurl a Shell Over 5,000 MPH". *Wired*.

[52] Osborn, Kris (2014-01-16). "Navy Rail Gun Showing Promise". *Defensetech.org*.

[53] Irwin, Sandra. "Naval Guns: Can They Deliver 'Affordable' Precision Strike?". *www.nationaldefensemagazine.com*. National Defense Magazine. Retrieved 11 February 2015.

[54] Sharp, David (2014-02-18). "US Navy Ready to Deploy Laser for 1st Time". *Military.com*.

[55] Atherton, Kelsey D. (2014-04-08). "The Navy Wants To Fire Its Ridiculously Strong Railgun From The Ocean". *Popular Science*.

[56] LaGrone, Sam (2013-06-07). "NAVSEA on Flight III Arleigh Burkes". *USNI News*.

[57] "Navy to deploy electromagnetic railgun". *Military1.com*. 2014-04-07.

[58] Bacon, Lance M. (2014-04-12). "Railgun tests from ship set for '16". *Military Times* (Gannett Government Media). Archived from the original on 2014-04-14. Retrieved 2014-12-22.

[59] Freedberg Jr., Sydney J. (2014-04-07). "Navy's Magnetic Super Gun To Make Mach 7 Shots At Sea In 2016: Adm. Greenert". *Breakingdefense.com*.

[60] "Electromagnetic Rail Gun EMRG". *www.globalsecurity.org*. Globalsecurity.org. Retrieved 10 February 2015.

[61] Frost, Tracy. "Survivable Electronics for Control of Hypersonic Projectiles under Extreme Acceleration". *http://www.navysbir.com*. United States Navy. Retrieved 10 February 2015. External link in |website= (help)

[62] Archived June 26, 2015, at the Wayback Machine.

[63] Witherspoon, F.D.; Bomgardner, R.; Case, A.; Messer, S.; et al. (2009). *MiniRailgun accelerator for plasma linear driven HEDP and magneto-inertial fusion experiments* (PDF). 36th IEEE International Conference on Plasma Science. San Diego, CA.

[64] Sam LaGrone (2015-04-14). "NAVSEA Details At Sea 2016 Railgun Test on JHSV Trenton - USNI News". News.usni.org. Retrieved 2015-12-24.

2.8.7   External links

- NRL Railgun Demonstration Video US Naval Research Laboratory, July 2010

- USN sets five-year target to develop electromagnetic gun at the Wayback Machine (archived November 22, 2009) Jane's Defence Weekly, 20 July 2006

- Electromagnetic Railgun Popular Science Article, June 2004

- Video of Navy railgun test firing, Navy Electromagnetic Launch Facility, Test Shot #1, 2 October 2006. Source: Fredericksburg.com, accessed 30 January 2007

- World's Most Powerful Rail Gun Delivered to Navy, 14 November 2007

Template:Non-rocket space launch

2.9   Stealth ship

USS Zumwalt *after floating out of drydock, 28 October 2013*

A **stealth ship** is a ship which employs stealth technology construction techniques in an effort to ensure that it is harder to detect by one or more of radar, visual, sonar, and infrared methods. These techniques borrow from stealth

U.S. Navy's Sea Shadow *(IX-529)*

aircraft technology, although some aspects such as wake and acoustic signature reduction (Acoustic quieting) are unique to stealth ships' design.

Reduction of radar cross-section (RCS), visibility and noise is not unique to stealth ships; visual masking has been employed for over two centuries and RCS reduction traces back to American and Soviet ships of the Cold War. One common feature is the inward-sloping tumblehome hull design that significantly reduces the RCS.

2.9.1   Examples

Several surface vessels employ stealth technology, amongst them the Russian *Admiral Gorshkov*-class frigate, the Swedish *Visby*-class corvette, the Dutch *Zeven Provinciën*-class frigate, the Turkish MİLGEM corvette, the Norwegian *Skjold*-class corvette, the French *La Fayette*-class frigate, the Chinese PLA Navy's Type 022 missile boat, Type 054A frigate, Type 056 corvette, the spanish *Álvaro de Bazán*-class frigate, Type 052C destroyer, Type 052D destroyer, and Type 055 destroyer, the German MEKO ships *Braunschweig*-class corvettes and *Sachsen*-class frigates, the Indian *Shivalik*-class frigate, *Kolkata*-class destroyer, Kamorta-class corvette, the Singaporean *Formidable*-class frigate, the British Type 45 destroyer, the U.S. Navy's *Zumwalt*-class destroyer, Finnish *Hamina*-class missile boats, Chilean Patrol Vessel PZM based on the German OPV80 and Indonesian 63m Stealth Fast Missile Patrol Vessel. Egypt has Ambassador MK III missile boat.

- HMS *Helsingborg*, one of the Swedish Navy's *Visby*-class corvettes

- French frigate *Surcouf* of the *La Fayette* class

- Dutch frigate *Evertsen* of the *Zeven Provinciën* class

- Indian stealth frigate INS *Shivalik*, lead ship of her class

Visby is designed to elude visual detection, radar detection, acoustic detection, and infrared detection. The hull material is a sandwich construction comprising a PVC core with a carbon fibre and vinyl laminate.[1] Avoidance of right angles in the design results in a smaller radar signature, reducing the ship's detection range.

Britain's Type 45 anti-air warfare destroyer has similarities to the *Visby* class, but is much more conventional, employing traditional steel instead of carbon fibre. Like *Visby*, its design reduces the use of right angles.

Sea Shadow, which utilizes both tumblehome and SWATH features, was an early U.S. exploration of stealth ship technology.

The currently developed U.S. *Zumwalt*-class destroyer — or DD(X) — is the US version of a stealth ship. Despite being 40% larger than an *Arleigh Burke*-class destroyer the radar signature is more akin to a fishing boat, according to a spokesman for Naval Sea Systems Command;[2] sound levels are compared to the *Los Angeles*-class submarines. The tumblehome hull reduces radar return and the composite material deckhouse also has a low radar return. Water sleeting along the sides, along with passive cool air induction in the mack reduces infrared signature.[3] Overall, the destroyer's angular build makes it "50 times harder to spot on radar than an ordinary destroyer.[2]

The *Arleigh Burke*-class destroyer also employs stealth technology without being a full stealth ship, similar to the German designs.

- German *Braunschweig*-class corvette

- Type 45 Royal Navy destroyer

- *Horizon*-class frigate of the French Navy and Italian Navy

Ady Gil, operated by the Sea Shepherd Conservation Society, was painted with what the owners claimed to be radar-absorbent material. Thus, *Ady Gil* would have been a rare case of a non-military vessel employing stealth technology.[4]

The Indonesian X3K 63m stealth fast missile patrol vessel has a trimaran design with long, but small center hull and has been built from carbon composite materials.[5]

2.9.2 Shaping

Main article: Naval architecture

In designing a ship with reduced radar signature, the main

Detail of the Forbin, *a modern frigate of the French navy. The faceted appearance reduces radar cross-section for stealth.*

Sloped surface features visible in this frontal view of Hamburg, *a* Sachsen-*class frigate of the German Navy*

concerns are radar beams originating near or slightly above the horizon (as seen from the ship) coming from distant patrol aircraft, other ships or sea-skimming anti-ship missiles with active radar seekers. Therefore, the shape of the ship avoids vertical surfaces, which would perfectly reflect any such beams directly back to the emitter. Retro-reflective right angles are eliminated to avoid causing the *cat's eye* effect. A stealthy ship shape can be achieved by constructing the hull and superstructure with a series of slightly protruding and retruding surfaces. This design was developed by several German shipyards, and is thus extensively applied on ships of the German Navy.

2.9.3 See also

- F125-class frigate
- *Steregushchy*-class corvette
- *Milgem*-class corvette
- *Visakhapatanam*-class destroyer
- *Kolkata*-class destroyer
- *Shivalik*-class frigate
- *Talwar*-class frigate
- *Kamorta*-class corvette
- *Brandenburg*-class frigate
- *Sigma*-class corvette
- *Sa'ar V*-class missile boat
- *Valour*-class frigate
- ARC *Juan Ricardo Oyola Vera*
- Infrared signature

2.9.4 References

[1] "Visby Class, Sweden". *www.naval-technology.com*. Retrieved 2015-07-31.

[2] Patterson, Thom; Lendon, Brad (14 June 2014). "Navy's stealth destroyer designed for the video gamer generation". CNN. Retrieved 29 October 2014.

[3] "DDG-1000 Zumwalt / DD(X) Multi-Mission Surface Combatant". GlobalSecurity.org. 1 September 2008.

[4] "Exclusive: The Earthrace Begins A Sea Shepherd Transformation « ecorazzi.com :: the latest in green gossip". Ecorazzi.com. Retrieved 2010-11-12.

[5] "KRI Klewang". August 31, 2012.

- "Stealth on the Water" "Mechanical Engineering"

2.9.5 External links

- DDG 1000 information; formerly DD(X)

- Shivalik class frigate

- Stealth ships steam ahead article by Chris Summers on BBC news, 10 June 2004

2.10 Tumblehome

Broadside on a model of a French 74-gun ship from 1755. The narrowing of the hull and reduced calibre of the artillery is clearly visible as one rises to the deck.

In naval architecture, the **tumblehome** is the narrowing of a ship's hull with greater distance above the water-line. Expressed more technically, it is present when the beam at the uppermost deck is less than the maximum beam of the vessel.

A small amount of tumblehome is normal in many designs in order to allow any small projections at deck level to clear wharves.[1]

2.10.1 Origins

Tumblehome was common on wooden warships for centuries. In the era of oared combat ships it was quite common, placing the oar ports as far abeam as possible. This also made it more difficult to board by force, as the ships would come to contact at their widest points, with the decks some distance apart. The narrowing of the hull above this point made the boat more stable by lowering the weight above the waterline, which is one of the reasons it remained common during the age of cannon-armed ships. In addition, the sloping sides of a ship with an extreme tumblehome (45 degrees or more) increased the effective thickness of the hull versus flat horizontal trajectory gunfire (a straight line through faced more material to penetrate) and increased the likelihood of a shell striking the hull being deflected—much the same reasons that later tank armor was sloped.

French battleship Jauréguiberry *of 1891, showing extreme tumblehome construction*

It can be seen as well in steel constructed warships of the early 1880s when the United States and most European navies began building steel warships. France was predominantly strong in promoting the tumblehome design in their warships, advocating tumblehome to reduce the weight of the upper deck, as well as making the vessel more seaworthy and creating greater freeboard.[2] France sold their newly constructed pre-dreadnought battleship *Tsesarevich* to the Russian Imperial Navy in time for it to fight as Admiral Wilgelm Vitgeft's flagship at the Battle of the Yellow Sea on 10 August 1904. The Russo-Japanese War proved that the tumblehome battleship design was excellent for long distance navigation, especially when encountering narrow canals, and other waterways; but that it could be dangerously unstable when watertight integrity was breached.[3] Four tumblehome *Borodino*-class battleships, which had been built in Russian yards to *Tsesarevich*'s basic design, fought on 27 May 1905 at Tsushima. The fact that three of the four were lost in this battle resulted in the discontinuing

of the tumblehome design in future warships for nearly all navies.

A degree of tumblehome also facilitates paddling in a canoe or kayak,[4] while a greater degree of flare (its opposite) accommodates more cargo.[5]

2.10.2 Modern warship design

Comparison of conventional hull and the Visby-class corvette

Tumblehome has been used in proposals for several modern ship projects. The hullform in combination with choice of materials and so forth results in decreased radar reflection, which together with other signature (sound, heat etc.) dampening measures makes stealth ships. This faceted appearance is common application of the principles of stealth aircraft. The US Navy's Zumwalt-class destroyers are a modern example.

Due to stability concerns, most warships with wave-piercing hulls combine tumblehome with multi-hull designs, such as the Type 022 missile boat.

2.10.3 In narrowboat design

Main article: Narrowboat

The inward slope of a narrowboat's superstructure (from gunwales to roof) is referred to as tumblehome. The amount of tumblehome is one of the key design choices when specifying a narrowboat, because the widest part of a narrowboat is rarely more than 7 feet across, so even a modest change to the slope of the cabin sides makes a significant difference to the "full-height" width of the cabin interior. Too great a tumblehome would make a boat difficult to pass through for a tall person; too little and the cabin roof edges are at risk of damage when the boat is passing through a tunnel (many canal tunnels on the British inland waterways have subsided, bringing the curve of the roof closer to the water level).

2.10.4   In automobile design

The inward slope of the "greenhouse" above the beltline is also called the tumblehome. Less commonly, the inward curve of the body near the bottom may also be called a tumblehome. In 21st century automobile designs this *turnunder* is less pronounced or eliminated to reduce aerodynamic drag and to help keep the lower portions of the vehicle cleaner under wet conditions.

2.10.5   In railway design

Tumblehome can be seen where the carriage body attaches to the underframe in this photo of a North British Railway 3rd Class carriage from around 1900

The inwardly curving portions of railway passenger carriages at the point where the carriage sides join the underframes is also called the tumblehome. Tumblehome styling of railway carriages was particularly prevalent in Britain and Ireland (or on railways influenced by British engineers or equipment builders) in the 19th century and "wood body" era of the early 20th century. This enabled the wooden step running the length of the carriage to remain within the dimensions of the loading gauge, while allowing maximum width for the main body of the carriage. Thus there was space to place a foot when entering or leaving the carriage.

A tumblehome remains a feature of railway carriages in Great Britain and can be seen in most modern designs of passenger rolling stock.

Some recent vehicle designs for continental Europe, such as the "Lint" and "Talent" vehicles, also feature a tumblehome profile, which in some vehicles leads to the need for a retractable step to bridge the gap between vehicle floor and station platforms. The operation time of these steps, which must be fully extended before the sliding doors may be opened, can be observed to increase the train waiting

time ("dwell time") at stations, compared to vehicles without such steps.

2.10.6   References

Footnotes

[1] Pursey p. 218.

[2] Forczyk, p. 18.

[3] Forczyk, p. 76.

[4] Mather, 1885.

[5] Vaillancourt.

Works cited

- Forczyk, Robert (2009).   *Russian Battleship vs Japanese Battleship, Yellow Sea 1904–05*. Osprey. ISBN 978-1-84603-330-8.

- Mather, Frederic G. (1885). *The Evolution of Canoeing*

- Pursey, H. J. (1959). *Merchant Ship Construction Especially Written for the Merchant Navy*

- Vaillancourt, Henri.   Traditional Birchbark Canoes Built in the Malecite, Penobscot and Passamaquoddy style

- DDG-1000 Zumwalt / DD(X) Multi-Mission Surface Combatant Future Surface Combatant. GlobalSecurity.org. Modern use of tumblehome.

2.11   United States naval gunfire support debate

The **United States naval gunfire support debate** is an ongoing debate among the United States Navy, Marine Corps, Congress, and independent groups like the *United States Naval Gunfire Support Association* over what role naval gunfire support and naval surface fire support (NSFS) should play within the navy and how such a role can best be provided. At the heart of the issue is the role that naval gunfire support—the use of naval artillery to provide fire support for amphibious assault and other troops operating within their range—should play in the U.S. Navy of the 21st century.

Although the debate at large traces its roots back to the end of World War II, the current debate began in 1992 with the

The battleship USS New Jersey *fires at positions near Beirut on 9 January 1984 during the Lebanese Civil War.*

USS Wisconsin, *shown moored in Norfolk, Virginia, is one of four* Iowa-*class battleships open to the public as museums, and was one of two maintained in the US Mothball fleet for potential reactivation.*

retirement of the last active *Iowa*-class battleship, USS *Missouri* (BB-63), as a result of the reduced demand for naval artillery, the rise of ship and submarine-launched missiles and aircraft-launched precision guided munitions (such as laser-guided bombs, which can accurately strike and destroy an enemy target with a single strike). The most striking point of the debate in the United States centers on battleships: owing to the longtime maintenance and upkeep that the four completed *Iowa*-class battleships have undergone during their time in the navy's active and mothball fleets, many still view battleships as viable solutions for gunfire support, and these members have questioned if the navy can adequately replace the gunfire support provided by a battleship's main guns with the smaller guns on its current fleet of cruisers and destroyers.

The debate has played out across a wide spectrum of media, including newspapers, magazines, web blogs, and congressional research arms like the Government Accountability Office. Each side has presented different arguments on the best approach to the problem, but most of the participants favor the continuation of the DD(X) program or the reinstatement of the *Iowa*-class battleships to the Naval Vessel Register. The *Iowa*-class battleships, the *Arleigh Burke*-class destroyers, and *Zumwalt*-class destroyers have entered the debate as options put forward for naval gunfire support, while others advocate the use of specifically designed close air support planes and newer missile systems that can loiter in an area as a replacement for naval gunfire.

2.11.1 Background

Within a few years of the end of World War II, the United States deactivated all of its remaining battleships and placed them in the United States Navy reserve fleets. Most of these ships were eventually scrapped, but the four *Iowa*-class battleships were not, and on several occasions one or more of

these four battleships were reactivated for naval gunfire support. The U.S. Navy has held onto the four *Iowa*-class battleships long after the upkeep and maintenance of operating and maintaining a battleship and the arrival of aircraft and precision guided munitions led other nations to scrap their big-gun fleets.[1] Congress was largely responsible for keeping the four *Iowa*-class battleships in the United States Navy reserve fleets and on the Naval Vessel Register as long as they did. The lawmakers argued that the battleships' large-caliber guns had a useful destructive power that is lacking in the smaller, cheaper, and faster guns mounted by U.S. cruisers and destroyers.[2]

In the 1980s, President Ronald Reagan proposed creating a 600-ship navy as part of the overall defense department build-up to counter the threat of the armed forces of the Soviet Union; both the Soviet Army and Navy had grown in the aftermath of the unification of Vietnam in 1975 and the loss of faith that Americans had in their armed services.[3] As part of this, all four *Iowa*-class battleships were modernized and reactivated. However, when the Soviet Union collapsed in 1991, the 600-ship navy was seen as too costly to maintain, and so the navy made plans to return to its traditional 313-ship fleet.[4][5] This led to the deactivation of many ships in the navy's fleet, including the four reactivated battleships; all were removed from service between 1990 and 1992.[6][7][8][9] Originally, the navy had struck all four ships and made plans to donate them, however Congress intervened in this plan with the passing of the National Defense Authorization Act of 1996. Section 1011 required the United States Navy to reinstate to the Naval Vessel Register two of the *Iowa*-class battleships that had been struck by the navy in 1995; these ships were to be maintained in the United States Navy Reserve Fleets. The Navy was to ensure that both of the reinstated battleships were in good condition and could be reactivated for use in the Marine Corps' amphibious operations. Both battleships were to be maintained with the reserve fleet until such a time as the navy could certify that it had within its fleet the operational capacity to meet or exceed the gunfire support that both battleships could provide.[10] To comply with this requirement, the navy selected the battleships *New Jersey* and *Wisconsin* for reinstatement to the Naval Vessel Register.

New Jersey remained in the mothball fleet until the Strom Thurmond National Defense Authorization Act of 1999 passed through the United States Congress 18 October 1998. Section 1011 required the United States Secretary of the Navy to list and maintain *Iowa* and *Wisconsin* on the Naval Vessel Register, while Section 1012 required the Secretary of the Navy to strike *New Jersey* from the Naval Vessel Register and transfer the battleship to a not-for-profit entity in accordance with section 7306 of Title 10, United States Code. Section 1012 also required the transferee to locate the battleship in the State of New Jersey.[11] The navy made the switch in January 1999.[10] *Iowa* and *Wisconsin* were finally stricken from the Naval Vessel Register in 2006.

2.11.2   Replacing the battleships

The navy sees the battleships as prohibitively expensive,[12] and is working to persuade Congress to allow it to remove *Iowa* and *Wisconsin* from the Naval Vessel Register by developing extended-range guided munitions and a new ship to fulfill marine corps requirements for naval surface fire support (NSFS).

Arleigh Burke-*class guided missile destroyers are equipped with a 5-inch (127 mm) gun. The USN is attempting to develop a shell for this gun that can reach out 40 nautical miles (70 km) inland or more.*

The navy plan originally called for the extension of the range of the 5-inch (127 mm) guns on the Flight I *Arleigh*

Burke-class guided missile destroyers (USS *Arleigh Burke* to *Ross*) with Extended Range Guided Munitions (ERGMs) that would enable the ships to fire precision guided projectiles about 40 nautical miles (70 km) inland. The program was initiated in 1996 with a preliminary cost of US $78.6 million; however, the cost of the program increased 400% during its research and development phase. The results of the program had been similarly disappointing: the original expected operational capability date was pushed from 2001 to 2011 before being cancelled by the navy in March 2008 for budget-related reasons and an apparent shift by the navy from the ERGM program to the Ballistic Trajectory Extended Range Munition (BTERM) program.[13][14] These weapons are neither intended nor expected to satisfy the full range of the marine corps requirements.[15]

The result of the latter effort to design and build a replacement ship for the two battleships was the *Zumwalt*-class destroyer program, also known either as the DD(X) or DDG-1000 (in reference to the hull number assigned to *Zumwalt*). The DD(X) was to mount a pair of Advanced Gun System turrets capable of firing specially designed Long Range Land Attack Projectiles some 60 miles (100 km) inland. Originally, the navy had planned to build a total of 32 of these destroyers, however the increasing cost of the program led the navy to reduce the overall number of destroyers built from 32 to 24.[16] In 2007 the total procurement of *Zumwalt*-class destroyers was further reduced to a total of seven, before being discontinued at a total of two destroyers in July 2008 as a result of the high cost of building each of the two ships.[17] In September 2008 the navy and the House of Representatives reached an agreement which will allow for the construction of a third DD(X) destroyer, bringing the total number of *Zumwalt*-class destroyers to three.[18]

The discontinuation of the class is due in part to concerns that the *Zumwalt* ships may deprive other projects of needed funding, a concern that has been raised by the Congressional Budget Office (CBO), Congressional Research Service (CRS), and the Government Accountability Office, all of which have issued reports that suggest that total cost of each ship could be as high as $5 billion or more.[17][19] In addition to the high cost, naval officials discussing the cancellation of the DD(X) program cited the inability of the DD(X) to fire the Standard missile or provide adequate air defense coverage, and a "classified threat" which the navy feels can be better handled by the current *Arleigh Burke*-class destroyers than by the *Zumwalt*-class destroyers.[20] The article also reported that the Marine Corps no longer needs the long-range fire support from the *Zumwalts'* 155 mm Advanced Gun System because such fire support can be provided by Tactical Tomahawk cruise missiles and precision airstrikes.[20]

2.11.3 Striking the *Iowa*-class battleships

"DDG 1000 *Zumwalt* is [...] being developed by the Navy to serve as the backbone of tomorrow's surface fleet. DDG 1000 *Zumwalt* provides a broad range of capabilities that are vital both to supporting the Global War on Terror and to fighting and winning major combatant operations. *Zumwalt*'s multi-mission warfighting capabilities are designed to counter not only the threats of today, but threats projected over the next decade as well."

Statement of the DD(X) program on the United States Navy's Program Executive Office, Ships[21]

On 17 March 2006, while the ERGM and DD(X) programs were under development, the Secretary of the Navy exercised his authority to strike *Iowa* and *Wisconsin* from the Naval Vessel Register, which cleared the way for both ships to be donated for use as museums. The United States Navy and the United States Marine Corps had both certified that battleships would not be needed in any future war, and have thus turned their attention to development and construction of the next generation *Zumwalt*-class guided missile destroyers.

However, this move has drawn fire from sources familiar with the subject; among them are dissenting members of the United States Marine Corps. These dissenters argue that battleships are still a viable solution to naval gunfire support,[22][23] members of the United States Congress who remain "deeply concerned" over the loss of naval surface gunfire support that the battleships provided,[13] and a number of independent groups such as the United States' Naval Fire Support Association (USNFSA) whose ranks frequently include former members of the armed service and fans of the battleships.[24][25] Although the arguments presented from each group differ, they all agree that the United States Navy has not in good faith considered the potential of reactivated battleships for use in the field, a position that is supported by a 1999 Government Accountability Office report regarding the United States Navy's gunfire support program.[15]

"In summary, the committee is concerned that the Navy has foregone the long-range fire support capability of the battleship, has given little cause for optimism with respect to meeting near-term developmental objectives, and appears unrealistic in planning to support expeditionary warfare in the mid-term. The committee views the Navy's strategy for providing naval surface fire support as 'high risk,' and will continue to monitor progress accordingly."

Evaluation of the United States Navy's naval surface fire support program in the National Defense Authorization Act of 2007[13]

In response, the navy has pointed to the cost of reactivating the two *Iowa*-class battleships to their decommissioned capability. The navy estimates costs in excess of $500 million,[26][27] but this does not include an additional $110 million needed to replenish the gunpowder for the 16-inch (406 mm) guns because a survey found the powder to be unsafe. In terms of schedule, the Navy's program management office estimates that reactivation would take 20 to 40 months, given the loss of corporate memory and the shipyard industrial base.[2]

Reactivating the battleships would require a wide range of battleship modernization improvements, according to the navy's program management office. At a minimum, these modernization improvements include command and control, communications, computers, and intelligence equipment; environmental protection (including ozone-depleting substances); a plastic-waste processor; pulper/shredder and wastewater alterations; firefighting/fire safety and women-at-sea alterations; a modernized sensor suite (air and surface search radar); and new combat and self-defense systems.[2] The navy's program management office also identified other issues that would strongly discourage the Navy from reactivating and modernizing the battleships. For example, personnel needed to operate the battleships would be extensive, and the skills needed may not be available or easily reconstituted.[28] Other issues include the age and unreliability of the battleships' propulsion systems and the fact that the navy no longer maintains the capability to manufacture their 16-inch (410 mm) gun system components and ordnance.[2]

Although the navy firmly believes in the capabilities of the DD(X) destroyer program, members of the United States Congress remain skeptical about the efficiency of the new destroyers when compared to the battleships.[15] Partially as a consequence, Congress passed Pub. L. 109-364, the National Defense Authorization Act 2007, requiring the battleships be kept and maintained in a state of readiness should they ever be needed again.[29] Congress has ordered that the following measures be implemented to ensure that, if need be, *Iowa* and *Wisconsin* can be returned to active duty:

1. *Iowa* and *Wisconsin* must not be altered in any way that would impair their military utility;

2. The battleships must be preserved in their present condition through the continued use of cathodic protection, dehumidification systems, and any other preservation methods as needed;

3. Spare parts and unique equipment such as the 16-inch (410 mm) gun barrels and projectiles be preserved in adequate numbers to support *Iowa* and *Wisconsin*, if reactivated;

4. The navy must prepare plans for the rapid reactivation of *Iowa* and *Wisconsin* should they be returned to the navy in the event of a national emergency.[29]

These four conditions closely mirror the original three conditions that the Nation Defense Authorization Act of 1996 laid out for the maintenance of *Iowa* and *Wisconsin* while they were in the Mothball Fleet.[4][10]

2.11.4   Alternatives to Naval Gunfire

During the period of time in which the battleships were out of commission in the United States, several technological updates and breakthroughs enabled naval ships, submarines, and aircraft to compensate for the absence of big guns within the fleet.

Air superiority

F4U-5 Corsairs provide close air support to U.S. marines fighting Chinese forces during the war in Korea, December 1950.

The earliest challenge to naval artillery was the advent of aircraft and armour piercing/incendiary bombs, which could be used against land based targets in support of troop formations ashore. Although in its infancy during and after World War I, some saw the potential for aircraft and sea based air support and envisioned the role it would have in future conflicts. Among the more notable individuals within the United States was Brigadier General Billy Mitchell. Mitchell had served in World War I, where he eventually commanded all U.S. aircraft in the war and was responsible for leading Allied aircraft in support of the ground offensive during the Battle of Saint-Mihiel, one of the first coordinated air-ground offensives in history. Mitchell's experience in World War I led him to believe that battleships were

out of date, and he became an increasingly vocal proponent of air power.[30]

In 1921, Mitchell first demonstrated to the world that battleships and other gun dependent vessels could be sunk by aircraft loaded with heavy bombs. In one of his most famous demonstrations, Mitchell convinced the Navy to allow bomb loaded aircraft to attack the German dreadnought *Ostfriesland*, a battleship taken as a prize of war by the United States in 1918. Although the Navy had placed strict rules on the bombing exercise, Mitchell and his men violated the rules and attacked the battleship head on, which caused the vessel to sink in a mere 22 minutes.[31] Although downplayed at the time this would have a dramatic effect on U.S. policy, leading to increased research and development for aircraft.[32]

By World War II naval aircraft had evolved to the point where they posed a threat to battleships and other naval vessels that lacked sufficient anti-aircraft defense. During WWII air raids accounted for the loss of warships and merchant vessels of all types, including the battleships *Conte di Cavour*, *Arizona*, *Utah*, *Oklahoma*, *Prince of Wales*, *Roma*, *Musashi*, *Tirpitz*, *Yamato*, *Schleswig-Holstein*, *Impero*, *Lemnos*, *Kilkis*, *Ise* and *Hyūga*. These losses were sustained even after the introduction of the "All or Nothing" armor scheme (armor belts intended to protect battleships from guns of an equal or lesser caliber than their own) and the recognition of the role of airpower and the rise of various ship based anti-aircraft guns meant to improve air defense aboard ships.[33][34] In addition to their role in attacking ships, several aircraft like the P-47 Thunderbolt were employed for close air support for ground based troops in Europe and in the Pacific.[35]

By the time of the Korean War air power had been supplemented by the introduction of the jet engine, which allowed fighter and bomber aircraft to fly faster. As with their World War II predecessors, the newer jet aircraft proved capable of providing close air support for ground based troops, and were instrumental in aiding UN ground forces during the Battle of Chosin Reservoir.[36][37]

The Vietnam War saw the introduction of helicopter gunships which could be employed to support ground based forces, and the experience gained in Vietnam would spawn the creation of several aircraft during and after the war designed specifically to aid ground forces, including the AC-47 Spooky, Fairchild AC-119, Lockheed AC-130, and A-10 Thunderbolt II, all of which are operated by the Air Force, and the F/A-18 Hornet which is operated by the navy. In addition, the army and marine corps operate UH-1 Iroquois, AH-1 Cobra, and AH-64 Apache helicopters for close air support, and these helicopters can be stationed onboard amphibious assault ships to provide ship-to-shore air support for ground forces. These aircraft would later prove instrumental in aiding ground forces from the 1980s onwards, and would be involved in the 1991 Gulf War, the 2001 invasion of Afghanistan, and the 2003 invasion of Iraq.

Starting after the invasion of Iraq, the air force began arming unmanned drone aircraft to perform strike missions. Originally designed for prolonged surveillance (and ironically to act as spotters for naval artillery), these aircraft typically have greater endurance than manned strike aircraft and some degree of automation to allow them to patrol for activity without requiring the constant attention of a pilot. This permitted the fielding of a less expensive aerial force which could maintain constant surveillance for enemy targets and conduct strikes on any targets encountered.

Missiles

Arleigh Burke-*class guided missile destroyers use their Vertical Launch Systems to fire Standard missiles during a live fire exercise*

Towards the end of World War II Germany introduced the V-1 cruise missile and V-2 ballistic missiles in combat against the Allied forces. The missiles arrived too late to alter the course of the war, but after the fall of Nazi Germany the V-1 and V-2 rockets would form the foundations for the space race and for the policy of Mutually Assured Destruction by providing each superpower with Ballistic Missiles and Submarine Launched Ballistic Missiles that could carry nuclear warheads.

The rise of precision strike munitions in the 1970s and 1980s reduced the need for a massive naval bombardment against an enemy force, as missiles could now be used against such targets to support ground forces and to destroy targets in advance of the arrival of troops. Guided missiles can also fire much further than the guns of any destroyer, cruiser, frigate, or battleship, allowing for strikes deep into the heart of enemy territory without risking the lives of pilots or airplanes. This led to a major shift in naval thinking, and as a result ships became more dependent on missile

magazines than on their guns for offensive and defensive capabilities. This was demonstrated in the 1980s, when all four recommissioned battleships were outfitted with missile magazines, and again in the 1991 Gulf War, when both *Missouri* and *Wisconsin* launched missile volleys against targets in Iraq before using their guns against Iraqi targets on the coast. The same conflict saw the first use of submarine-launched cruise missiles when the *Los Angeles*-class attack submarine *Louisville* fired Tomahawk Land Attack Missiles into Iraq from the Red Sea.[38]

Currently, the United States is looking into Non-Line-of-Sight Launch Systems, which would fire either Precision Attack Munitions or Loitering Attack Munitions;[39][40] however the latter program has been cancelled due to rising costs and poor test performance,[41] while the Precision Attack Missile lacks the minimum range to meet the USMC requirement of 41.3 nautical miles (76.5 km).[42]

Although ship-fired missiles can provide support for shore-based units, they are susceptible to interception by anti-missile systems such as the Aegis Combat System and MIM-104 Patriot system developed by the United States and used by NATO nations. These systems were designed to track and destroy both artillery shells and missiles. The first widely reported instances of such systems working came in 1991 when the US Patriot and Royal Navy Sea Dart missile system successfully intercepted and destroyed Iraqi Scud and Silkworm missiles.[38][43][44][45]

Gun support

Test firing a railgun at the Naval Surface Warfare Center, January 2008

Naval gunfire has been used intermittently since the end of the Second World War. By and large, the guns are small caliber guns found on modern frigates, cruisers, destroyers. The reason the *Iowa*-class battleships were maintained and used is because 16-inch (410 mm) guns were considered more effective.

In the 1960s, following a requirement established by Chief of Naval Operations (CNO) for a new gun capable of firing semi-active laser guided projectiles (SAL GP), the Naval Surface Warfare Center Dahlgren Division worked on the Major Caliber Lightweight Gun (MCLWG) program, testing capability of destroyer-sized ships to provide shore bombardment support with the range previously available from decommissioned cruisers. The 8"/55 caliber Mark 71 gun, a single gun version of the 8"/55 Mark 16 caliber gun was mounted aboard the USS *Hull* (DD-945). However after at-sea technical evaluation in 1975 and operational testing that followed through 1976, The Operational Test and Evaluation Force determined inaccuracy made the gun operationally unsuitable. The lightweight 8"/55 was concluded to be no more effective than the 5"/54 with Rocket Assisted Projectiles. Program funding was terminated in 1978.[46]

In the 1980s, such guns were used by US destroyers during the Lebanese Civil War to shell positions for the Multinational Force in Lebanon operating on the ground. Guns were also used by the Royal Navy in the Falklands War to support British forces during the operations to recapture the islands from the Argentinans. For example, the Type 42 destroyer HMS *Cardiff* was required to fire at enemy positions on the islands with her 4.5-inch gun. In one engagement she fired 277 high-explosive rounds,[47] although later problems with the gun prevented continual use. Ship-based gunfire was also used during Operation Praying Mantis in 1988 to neutralize Iranian gun emplacements on oil platforms in the Persian Gulf.[48] Although the smaller caliber guns are effective in combat, larger caliber guns can be employed for psychological warfare purposes, and have compelled the surrender of enemy combatants during combat operations due to a sense of overwhelming firepower. One of the most recent examples of this was the bombardment of Iraqi shore defenses by the battleships *Missouri* and *Wisconsin* in the Persian Gulf War.[38] The shelling proved to be so devastating that when the latter battleship returned to resume shelling the island, the enemy troops surrendered to her Pioneer UAV launched to spot for the battleships' guns rather than face another round of heavy naval artillery support.[49][50]

The navy has looked into creating precision guided artillery rounds for use with the current fleet of cruisers and destroyers. The most recent attempt to modify the guns for longer range came with the Advanced Gun System mounts that were to be installed aboard the *Zumwalt*-class destroyers, although the navy has been involved in the Long Range Land Attack Projectile and Ballistic Trajectory Extended Range Munition projects for over 10 years in an effort to develop Extended Range Guided Munitions.[13][14]

In addition to funding research into various extended range munitions, the navy is also working on developing railguns for use with the fleet at some point in the future. The United States Naval Surface Warfare Center Dahlgren Division demonstrated an 8 MJ rail gun firing 3.2 kilogram (slightly more than 7 pounds) projectiles in October 2006 as a prototype of a 64 MJ weapon to be deployed aboard navy warships. The main problem the navy has had with implementing a railgun cannon system is that the guns wear out due to the immense heat produced by firing. Such weapons are expected to be powerful enough to do a little more damage than a BGM-109 Tomahawk missile at a fraction of the projectile cost.[51] Since then, BAE Systems has delivered a 32 MJ prototype to the Navy.[52] On January 31, 2008, the US Navy tested a magnetic railgun; it fired a shell at 2520 m/s using 10.64 megajoules of energy.[53] Its expected performance is over 5800 m/s muzzle velocity, accurate enough to hit a 5 meter target from 200 nmi (370 km) away while shooting at 10 shots per minute. It is expected to be ready between 2020 and 2025.

Apart from railguns, 16 inch scramjet rounds with ranges of up to 400 nautical miles that have a 9-minute time of flight are being proposed by Pratt and Whitney working with Dr. Dennis Reilly, a plasma physicist with extensive experience with munitions. Alliant Techniques is also developing a ram-jet projectile for 5-inch and 155mm gun. Unfortunately, the navy had no interested sponsor according to both Pratt and Whitney representatives and Dr. Reilly.[54]

2.11.5   Recent developments

The Zumwalt*-class destroyers, also known either as DD(X) or DDG-1000, were to be the replacement ships for the battleships*

Prior to the reduction of ships in the DD(X) destroyer program, it seemed unlikely that the above four conditions would have impeded the current plan to turn *Iowa* and *Wisconsin* into museum ships because the navy had expected a sufficient number of DD(X) destroyers to be ready to help fill the NSFS gap by 2018 at the earliest;[2] however, the

July 2008 decision by the navy to cancel the DD(X) program would leave the navy without a ship class capable of replacing the two battleships removed from the Naval Vessel Register in March 2006. Although unlikely, the cancellation of the DD(X) destroyer program may result in a reinstatement of *Iowa* and *Wisconsin* to the Naval Vessel Register; by law, the navy is required to maintain two battleships on the register until the navy certifies that it has within its fleet the operation NSFS capability that can meet or exceed that provided by the battleships,[10] and with the Extended Range Guided Munitions program already cancelled in March 2008[14] and DD(X) destroyer program essentially cancelled in July 2008[17] the navy does not appear to have met its needed criteria for battleship removal.[10] James T. Conway, Commandant of the Marine Corps has said that missiles fired from the Littoral combat ship could fulfill the USMC needs for NSFS.[55] This would not be the current NLOS-LS program as the range of the PAM missile at 22 miles (35 km) falls short of the threshold requirement for NSFS of 41 miles (66 km) and the number of CLUs the current LCS designs can carry in a ready to fire configuration is also short of the required volume of fire.[56] The Loitering Attack Missile could have matched the required range, but it was cancelled in 2011[41] and the LCS would still have fallen short in terms of rounds ready to fire.[57]

On September 15th 2015, Republican presidential candidate, Donald Trump, while giving a speech on defense on board the battleship, USS *Iowa* (BB-61) in San Pedro, California, briefly remarked in having interest in recommissioning the *Iowa*-class battleships.[58]

2.11.6 Notes

[1] Government Accountability Office, *Naval Surface Fire Support Program Plans and Costs* (NSIAD-99-91).

[2] Government Accountability Office. *Information on Options for Naval Surface Fire Support* (GAO-05-39R).

[3] Holland, W. J. (2004). *The Navy*. China: Barnes & Noble, Inc., by arrangement with Hugh Lauter Levin Associates, Inc. p. 184. ISBN 076076218X.

[4] "BB-61 IOWA-class (Specifications)". Federation of American Scientists. Retrieved 2006-11-26.

[5] Johnston, Ian & McAuley, Rob (2002). *The Battleships*. London: Channel 4 Books (an imprint of Pan Macmillian, LTD). p. 183. ISBN 0-7522-6188-6.

[6] Naval Historical Center. *"Iowa"*. *Dictionary of American Naval Fighting Ships*. Navy Department, Naval History & Heritage Command. Retrieved 2008-10-07.

[7] Naval Historical Center. *"New Jersey"*. *Dictionary of American Naval Fighting Ships*. Navy Department, Naval History & Heritage Command. Retrieved 2008-10-07.

[8] Naval Historical Center. *"Missouri"*. *Dictionary of American Naval Fighting Ships*. Navy Department, Naval History & Heritage Command. Retrieved 2008-10-07.

[9] Naval Historical Center. *"Wisconsin"*. *Dictionary of American Naval Fighting Ships*. Navy Department, Naval History & Heritage Command. Retrieved 2008-10-07.

[10] 104th Congress, House of Representatives. National Defense Authorization Act of 1996. p. 237. Retrieved 17 December 2006.

[11] "Strom Thurmond National Defense Authorization Act of 1999 (Subtitle B-Naval Vessels and Shipyards)" (PDF). 105th Congress, United States Senate and House of Representatives. pp. 200–201. Retrieved 2007-03-12.

[12] John Pike. "BB-61 Iowa-class Reactivations". Globalsecurity.org. Retrieved 2010-02-06.

[13] "National Defense Authorization Act of 2007" (PDF). pp. 193–194. Retrieved 2007-03-12.

[14] Matthews, William (2007-03-25). "Navy ends ERGM funding". *Navy Times*. Retrieved 2008-04-23.

[15] Government Accountability Office, *Evaluation of the Navy's 1999 Naval Surface Fire Support Assessment* (NSAID-99-225).

[16] "National Defense Authorization Act of 2007" (pdf) pp. 109th Congress, United States Senate and House of Representatives. 69–70. Retrieved on 2008-08-01.

[17] Cavas, Christopher P. (2008-07-24). "DDG 1000 program will end at 2 ships". *Navy Times*. Retrieved 2008-07-27.

[18] "US House, Senate Agree to Add 3rd DDG 1000". Defense News. 24 September 2008. Retrieved 2008-10-07.

[19] Labs, Eric J. (2008-07-31). "The Navy's Surface Combatant Programs" (PDF). Congressional Budget Office. pp. 3–9. Retrieved 2008-08-02.

[20] Ewing, Philip; Bryan Mitchell (2008-08-01). "Navy:No Need to Add DDG 1000s After All". *defense news* (Army Times Publishing Company). Retrieved 2008-08-06.

[21] Program Executive Office, Ships (2007-05-27). "DDG 1000 (subsection: What is DDG 1000?)". United States Navy. Retrieved 2007-06-24.

[22] Novak, Robert (2005-12-06). "Losing the battleships". *CNN.com*. Retrieved 2008-07-25.

[23] The Marine Corps supports the strategic purpose of reactivating two battleships in accordance with the National Defense Authorization Act of 1996 and supports the Navy's modernization efforts to deliver a sufficient NSFS capability that exceeds that of the *Iowa*-class battleships. See: Government Accountability Office. *Information on Options for Naval Surface Fire Support*.

[24] Blazar, Ernest (1996-07-29). "New debate resurrects old one; critics say cancel arsenal ship, bring back battleships". *Navy Times*.

[25] "Navy proposes destroyer with long-range guns". *USA Today*. 2005-08-15.

[26] This number is based on 1999 estimate with a 4% annual inflation rate. See: Government Accountability Office. *Information on Options for Naval Surface Fire Support*.

[27] The U.S. Navy reported in the April 1987 edition of *All Hands* that the original cost of bringing the battleships back in the 1980s was $110 million per ship, but the actual cost after modernization and recommissioning was $455 million. See: Bureau of Naval Personnel, "Back on the battle line".

[28] The U.S. Navy reported in the April 1987 edition of *All Hands* that while battleships have larger crews than other vessels the level of training required and the criticality of that training were less than that required of a crew aboard an *Oliver Hazard Perry*-class frigate. See: Bureau of Naval Personnel, "Back on the battle line".

[29] 109th Congress, House of Representatives. Report 109–452. National Defense Authorization Act of 2007. p. 68. Retrieved 26 November 2006.

[30] Mitchell, William. *Winged Defense: The Development and Possibilities of Modern Air Power—Economic and Military*, p. 119. Dover Publications, 2006. ISBN 0-486-45318-9

[31] "Vice Admiral Alfred Wilkinson Johnson, USN Ret. "The Naval Bombing Experiments: Bombing Operations" (1959)". History.navy.mil. Retrieved 2010-02-06.

[32] Reid, John Alden. *Bomb the Dread Noughts!* Air Classics, 2006.

[33] Keegan, p. 264.

[34] Toppan, Andrew (2001-10-06). "World Battleships List: US Treaty and Post-Treaty Battleships". Retrieved 2007-06-01.

[35] "The Republic P-47 Thunderbolt". Greg Goebel. 2006-06-01. Retrieved 2008-10-07.

[36] National Museum of the USAF - Fact Sheet Media (F-86A/E/F Sabre)

[37] "American Military Aircraft (F-86 in Korea)". Home.att.net. Retrieved 2010-02-06.

[38] United States. Office of the Chief of Naval Operations. (1991-05-15). "V: "Thunder And Lightning"- The War With Iraq". *The United States Navy in "Desert Shield" / "Desert Storm"*. Washington, D.C.: United States Navy. OCLC 25081170. Retrieved 2006-11-26.

[39] Corrin, Amber (2010-05-14). "Embattled missile program meets its demise". Defensesystems.com. Retrieved 2011-05-27.

[40] "Non-Line-of-Sight Launch System (NLOS-LS)". globalsecurity.org. 2006-01-10. Retrieved 2008-10-09.

[41] "Gates Reveals Budget Efficiencies, Reinvestment Possibilities.". Defense.gov. Retrieved 2011-05-27.

[42] "Naval Surface Fire Support: Navy's Near-Term Plan Is Not Based on Sufficient Analysis (Letter Report, 05/19/95, GAO/NSIAD-95-160)".

[43] Lewis Page (27 November 2007). "New BAE destroyer launches today on the Clyde". The Register. Retrieved 2008-04-21.

[44] Bernard Rostker (19 September 2000). "TAB H -- Friendly-fire Incidents". United States Department of Defense. Retrieved 2008-08-11.

[45] The success of the MIM-104 Patriot Missile System in these engagements, and in particular how many of them were real targets is still controversial. Post war video analysis of presumed interceptions suggests that no Scud was actually hit. "Optical Evidence Indicating Patriot High Miss Rates During the Gulf War". Retrieved 2008-01-29.

[46] Miller, John C., Col USMC & Peterson, H.W., Major USMC "Guns vs. Butter - Without the Guns?" United States Naval Institute Proceedings January 1982 pp.33–34

[47] "Report of Proceedings". HMS *Cardiff*—The 1982 Ship's Company. Retrieved 2008-02-12.

[48] "Operation Earnest Will". globalsecurity.org. 2005-04-27. Retrieved 2008-10-07.

[49] Federation of American Scientists. Pioneer Short Range (SR) UAV. Retrieved 26 November 2006.

[50] National Air and Space Museum, Smithsonian Institution. Pioneer RQ-2A. 14 September 2001. Retrieved 26 November 2006.

[51] Zitz, Michael (2007-01-17). "A missile punch at bullet prices". Fredericksburg.com. Retrieved 2008-10-07.

[52] Sofge, Erik (2007-11-14). "World's Most Powerful Rail Gun Delivered to Navy". Popular Mechanics. Retrieved 2008-10-07.

[53] "U.S. Navy Demonstrates World's Most Powerful EMRG at 10 Megajoules".

[54] Shawn Welch, Colonel, Corps of Engineers United States Army. Joint and Interdependent Requirements: A Case Study in Solving the Naval Surface Fire Support Capabilities Gap. Joint Advanced Warfighting School Masters Thesis. Retrieved 2012-11-19 from Defense Technical Information Center.

[55] [http://defense.iwpnewsstand.com/websearch.asp?f= &docnum=NAVY-22-2-3&DOCID=CONWAY-+ ROCKETS+ON+LCS+COULD+FILL+NAVAL+ SURFACE+FIRES+REQUIREMENT+(NAVY-22-2-3)> "Conway: Rockets on LCS Could Fill Naval Surface Fires Requirement"]. Retrieved 2009-01-23.

[56] "Marines Pushing Ahead On Corps-Specific Module Ideas For LCS INSIDE THE NAVY 27 OCT 08". Archived from Documents/PAO Folder/MIW NEWS CLIPS/News Clips October 2008/0810-29.htm the original Check |url= value (help) on December 14, 2009. Retrieved 2009-01-24.

[57] "Commandant says Marines will field their own gunships". Govexec.com. Retrieved 2010-02-06.

[58] http://foxtrotalpha.jalopnik.com/ trump-s-crazy-idea-to-bring-back-battleships-might-actu-1731148

2.11.7 References

- Bureau of Naval Personnel (April 1987). "Back on the battle line" (PDF). *All Hands* (Washington, D.C.: United States Navy) **841**: 28–29. ISSN 0002-5577. OCLC 2555618. Retrieved 2008-06-27.

- Coram, Robert. *Boyd: The Fighter Pilot Who Changed the Art of War*. Los Angeles: Back Bay Books, 2004. ISBN 0-316-79688-3.

- DiGiulian, Tony (2008-07-02). "United States of America 16"/50 (40.6 cm) Mark 7". NavWeaps.com. Retrieved 2007-01-16.

- Government Accountability Office (1999-06-11). "Naval Surface Fire Support Program Plans and Costs (GAO/NSIAD-99-91)" (PDF). Government Accountability Office. Retrieved 2007-03-12.

- Government Accountability Office (1999-09-14). "Evaluation of the Navy's 1999 Naval Surface Fire Support Assessment (GAO/NSIAD-99-225)" (PDF). Government Accountability Office. Retrieved 2007-03-12.

- Government Accountability Office (2004-11-19). "Information on Options for Naval Surface Fire Support (GAO-05-39R)" (PDF). Government Accountability Office. Retrieved 2007-03-14.

- Major B. T. Kowalski, United States Marine Corps (1995). "Naval Surface Fire Support, Is It A Viable Option?".

- Keegan, John (2000). *World War II: A Visual Encyclopedia*. London: PRC Publishing. ISBN 1-85585-878-9. OCLC 45188820.

- Miller, David; Chris Miller (1986). *Modern Naval Combat*. London: Salamander Books. ISBN 0-86101-231-3. OCLC 17397400.

- Muir, Malcolm, Jr. (1987). *The Iowa-Class Battleships: Iowa, New Jersey, Missouri, Wisconsin*. Poole, Dorset: Blandford Press. ISBN 0-7137-1732-7. OCLC 17509226.

- Sumrall, Robert F. (1988). *Iowa Class Battleships: Their Design, Weapons & Equipment*. Annapolis, Maryland: Naval Institute Press. ISBN 0-87021-298-2. OCLC 19282922.

2.12 Vertical launching system

For the Brazilian Space Agency launcher, see VLS-1.

A **vertical launching system** (**VLS**) is an advanced sys-

The VLS cells on board USS San Jacinto

tem for holding and firing missiles on mobile naval platforms, such as surface ships and submarines. Each vertical launch system consists of a number of *cells*, which can hold one or more missiles ready for firing. Typically, each cell can hold a number of different types of missiles, allowing the ship flexibility to load the best set for any given mission. Further, when new missiles are developed, they are typically fitted to the existing vertical launch systems of that nation, allowing existing ships to use new types of missiles without expensive rework. When the command is given, the missile flies straight up long enough to clear the cell and the ship, and then turns on course.

A VLS allows surface combatants to have a greater number of weapons ready for firing at any given time compared to older launching systems such as the Mark 13 single-arm and Mark 26 twin-arm launchers, which were fed from behind by a magazine below the main deck. In addition to greater firepower, VLS is much more damage tolerant and reliable than the previous systems, and has a lower radar cross-section (RCS). The U.S. Navy now relies exclusively on VLS for its guided missile destroyers and cruisers.

The most widespread vertical launch system in the world is the Mark 41, developed by the US Navy. More than 11,000 Mark 41 VLS missile cells have been delivered, or are on order, for use on 186 ships across 19 ship classes, in 11 navies around the world. This system currently serves with the US Navy as well as the Australian, Canadian, Danish,

A Tomahawk missile canister being loaded into a VLS aboard the Arleigh Burke-*class destroyer USS* Curtis Wilbur

Dutch, German, Japanese, New Zealand, Norwegian, South Korean, Spanish, and Turkish navies, while others like the Greek Navy preferred the similar Mark 48 system.[1]

The advanced Mark 57 vertical launch system is projected to be used on the new *Zumwalt*-class destroyer. The older Mark 13 and Mark 26 systems remain in service on ships that were sold to other countries such as Taiwan and Poland.

When installed on an SSN (nuclear-powered attack submarine), a VLS allows a greater number and variety of weapons to be deployed in comparison to using only torpedo tubes.

2.12.1   Hot launch and cold launch

A vertical launch system can be either *hot launch*, where the missile ignites in the cell, or *cold launch*, where the missile is expelled by gas produced by a gas generator which is not part of the missile itself, and then the missile ignites. "Cold" means *relatively* cold compared with rocket engine exhaust. A hot launch system does not require an ejection

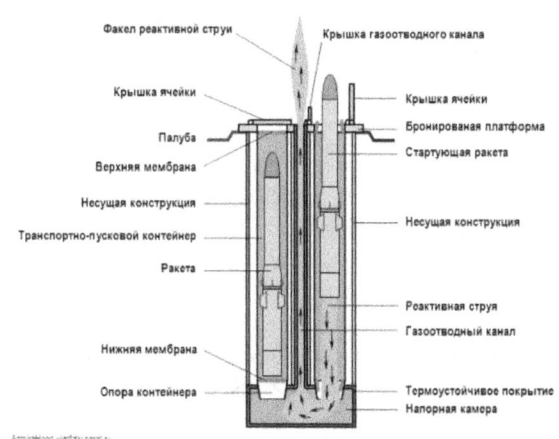

Diagram depicting a hot launch from a Mark 41 VLS

mechanism, but does require some way of disposing of the missile's exhaust and heat as it leaves the cell. If the missile ignites in a cell without an ejection mechanism, the cell must withstand the tremendous heat generated without igniting the missiles in the adjacent cells.

US Navy Mark 41 Tomahawk launch, hot launch.

An advantage of a hot-launch system is that the missile propels itself out of the launching cell using its own engine, which eliminates the need for a separate system to eject the missile from the launching tube. This potentially makes a hot-launch system relatively light, small, and economical to develop and produce, particularly when designed around smaller missiles. A potential disadvantage is that a malfunctioning missile could destroy the launch tube.

The advantage of the cold-launch system is in its safety: should a missile engine malfunction during launch, the cold-launch system can eject the missile thereby reducing or eliminating the threat. For this reason, Russian VLSs are often designed with a slant so that a malfunctioning missile will land in the water instead of on the ship's deck. As missile size grows, the benefits of ejection launching increase.

Above a certain size, a missile booster cannot be safely ignited within the confines of a ship's hull. Most modern ICBMs and SLBMs are cold-launched.

American surface-ship VLSs have the missile cells arranged in a grid with one lid per cell and are "hot launch" systems; the engine ignites within the cell during the launch, and thus it requires exhaust piping for the missile flames and gasses. France, Italy and Britain use a similar hot-launching Sylver system in PAAMS. Russia produces both grid systems and a revolver design with more than one missile per lid. Russia also uses a cold launch system for some of its vertical launch missile systems, e.g., the Tor missile system. The People's Republic of China uses a circular "cold launch" system that ejects the missile from the launch tube before igniting the engine on the Type 052C destroyer, and also a rectangular "hot launch" system with one lid per cell arranged in a grid on the Type 054A frigate.

2.12.2 Systems in use by nations

Australia

- Adelaide class frigate - Mark 41 (8 cells)
- Anzac class frigate - Mark 41 (8 cells)

Belgium

- Karel Doorman class frigate - Mark 48 Mod 1 (16 cells)

Canada

- Iroquois class destroyer - Mark 41 (29 cells)
- Halifax class frigate - Mark 48 Mod 0 (16 cells)

Chile

- Karel Doorman class frigate - Mark 48 Mod 1 (16 cells)

People's Republic of China

Surface

- Type 052D destroyer - (64 cells)

- Type 052C destroyer - HHQ-9 (48 cells)
- Type 051C destroyer - 48N6E (48 cells)
- Type 054A frigate - HQ-16 or Y-8 (32 cells)

Submarine

- Type 032 submarine - (4 cells)
- Type 093B submarine

Denmark

- Iver Huitfeldt class frigate - Mark 41 (32 cells) and Mark 56 (2 x 12 cells)
- Absalon class support ship- Mark 56 (3 x 12 cells)

France

- Charles de Gaulle (R91) aircraft carrier - SYLVER (32 cells)
- Horizon class frigate - SYLVER (48 cells)
- Aquitaine class FREMM multipurpose frigate - SYLVER (32 cells)

Germany

- Sachsen class frigate - Mark 41 (32 cells)
- Brandenburg class frigate - Mark 41 (16 cells)

Greece

- Hydra class frigate - Mark 48 (16 cells)

India

- INS Viraat - Barak 1 (16 cells)
- INS Vikramaditya - Barak 1 (24 cells) and Barak 8
- Kolkata class destroyer - Barak 8/Barak 1 (32 cells) and BrahMos (16 cells)
- Delhi class destroyer - Barak 1 (32 cells)

- Rajput class destroyer - BrahMos (8 cells) and Barak 1

- Shivalik class frigate - Club or BrahMos (8 cells) and Barak 1 (32 cells)

- Talwar class frigate - Club or BrahMos (8 cells)

- Brahmaputra class frigate - Barak 1 (24 cells)

- Godavari class frigate - Barak 1 (24 cells)

- Kamorta class corvette - Barak 1 (16 cells)

Indonesia

- Bung Tomo class corvette - VL MICA

- Van Speijk class frigate - Yakhont VLS (4 cells)

SYLVER cells of Italian destroyer Caio Duilio

Israel

- Sa'ar 5-class corvette - Barak 1 (2 x 32 cells)

Italy

- Cavour (550) aircraft carrier - SYLVER A43 (32 cells)

- Horizon class frigate - SYLVER A50 (48 cells)

- FREMM multipurpose frigate - SYLVER A50 (16 cells)

Japan

- Atago class destroyer - Mark 41 (96 cells)

- Kongō class destroyer - Mark 41 (90 cells)

- Hyūga class helicopter destroyer - Mark 41 (16 cells)

- Akizuki class destroyer (2010) - Mark 41 (32 cells)

- Takanami class destroyer - Mark 41 (32 cells)

- Murasame class destroyer - Mark 41 (16 cells) + Mark 48 (16 cells)

New Zealand

- Anzac class frigate - Mark 41 (8 cells)

Republic of Korea

- Gwanggaeto the Great class destroyer (KDX-I) - Mark 48 (16 cells)

- Chungmugong Yi Sun-shin class destroyer (KDX-II) - Mark 41 (32 cells) + K-VLS (24 cells)

- King Sejong the Great class destroyer (KDX-III) - Mark 41 (80 cells) + K-VLS (48 cells)

- Nampo-class minelayer - K-VLS (4 cells)

Netherlands

- De Zeven Provinciën class frigate - Mark 41 (40 cells)

- Karel Doorman class frigate - Mark 48 Mod 1 (16 cells)

Norway

- Fridtjof Nansen class frigate - Mark 41 (8 or 16 cells)

Soviet missile cruiser Frunze firing a missile from the Tor VLS

Top view of the Ticonderoga-*class USS* Lake Champlain *(CG-57) with VLS visible fore and aft as the gray boxes near the bow and stern of the ship.*

Portugal

- Karel Doorman class frigate - Mark 48 Mod 1 (16 cells)

South Africa

- Valour Class Frigate - Umkhonto (32 cells)

Russia

- Slava class cruiser - Fort (64 cells)
- Kirov class battlecruiser - Fort/Fort-M (96 cells) + Tor (128 cells)
- Udaloy class destroyer Tor (64 cells)

- Admiral Grigorovich class frigate - 3S14 Agat for Kalibr or BrahMos (8 cells)
- Admiral Gorshkov class frigate - 3S14 Agat for Kalibr or Oniks (8 cells)
- Neustrashimyy class frigate - Tor (4 x 8 cells)
- Buyan-M class corvette - 3S14 Agat for Kalibr or Oniks (8 cells)
- Gepard-class frigate - 3S14 Agat for Kalibr or Uran (8 cells)
- Steregushchy class corvette - 3S14 Agat for Kalibr (1 x 6 cells) or Oniks (2 x 4 cells) or Uran (1 x 8 cells)
- Gremyashchy class corvette - 3S14 Agat for Kalibr or Oniks (1 x 8 cells)

Singapore

- Formidable class frigate - SYLVER (32 cells)
- Victory class corvette - Barak 1 (2 x 8 cells)

Spain

- Álvaro de Bazán class frigate - Mark 41 (48 cells)

Thailand

- Naresuan class frigate - Mark 41 (8 cells)

Turkey

- G class frigate- 8 cell VLS
- Barbaros class frigate- VLS Mark 41 Mod 8 [2]

United Kingdom

Surface

- Type 45 destroyer - SYLVER (48 cells)
- Type 23 frigate - GWS.26 (32 cells)

Submarine

- Vanguard class - Ballistic Missile Tubes (B.M.T) (16 Cell)

🇺🇸 United States

Surface

- Arleigh Burke class destroyer - Mark 41 (96 cells)

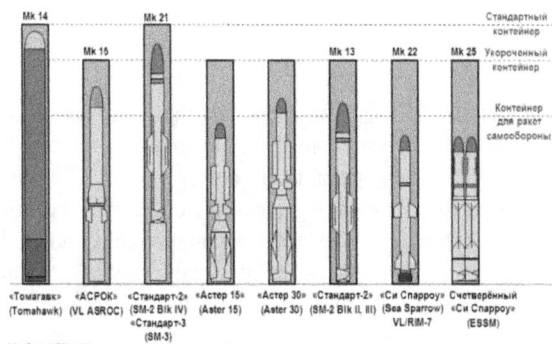

VLS Mark 41 Canister Types

- Ticonderoga class cruiser - Mark 41 (122 cells)
- Zumwalt class destroyer - Mark 57 (80 cells)

Submarine

- *Los Angeles*-class submarine - Mark 45 (12 tubes) for Tomahawk cruise missiles
- *Virginia*-class submarine - Mark 45 (12 tubes) for Tomahawk cruise missiles
- *Ohio*-class submarine, SSGN conversion - Mark 45 (154 tubes) for Tomahawk cruise missiles. 22 ballistic missile tubes were converted to allow for seven conventional VLS tubes in place of a SLBM tube.

2.12.3 See also

- List of United States Navy Guided Missile Launching System

2.12.4 References

[1] http://www.defenseindustrydaily.com/MK-41-Naval-Vertical-Missile-Launch-Systems-Delivered-Supported-updated-02139/#more-2139

[2] http://www2.ssm.gov.tr/katalog2007/data/09304/uruneng/4.htm

2.12.5 External links

- Mk 41 VLS - Federation of American Scientists
- MK 41 Vertical Launching System (VLS) - GlobalSecurity.org
- Mk-48 Vertical Launching System (VLS) - Seaforces-online Naval Information

Chapter 3

Zumwalt-Class Destroyers

3.1 USS Lyndon B. Johnson (DDG-1002)

USS *Lyndon B. Johnson* (**DDG-1002**) is to be the third and final ship of the *Zumwalt*-class destroyer. The contract to build her was awarded to Bath Iron Works located in Bath, Maine, on 15 September 2011. The award, along with funds for the construction of USS *Michael Monsoor*, was worth US$1.826 billion.[1][5] On 16 April 2012, Secretary of the Navy Ray Mabus announced the ship would be named *Lyndon B. Johnson* in honor of Lyndon B. Johnson, who served as the 36th President of the United States from 1963 to 1969. DDG-1002 is the 34th ship named by the Navy after a U.S. president.[6]

Lyndon B. Johnson is a *Zumwalt*-class destroyer, 32 units of which were originally planned, the U.S. Navy eventually reduced this number to three units.[7] Designed as multimission ships with an emphasis on land attack and littoral warfare,[8] the class features the tumblehome hull form, reminiscent of ironclad warships.[9] In January 2013 the Navy solicited bids for a steel deckhouse as an option for *Lyndon B. Johnson* instead of the composite structures of the other ships in the class.[10] This change was made in response to cost overruns for the composite structure, but due to the tight weight margins in the class, required weight savings in other parts of the ship.[11]

In February 2015, the Navy revealed they had begun engineering studies to include an electromagnetic railgun on *Lyndon B. Johnson*. The *Zumwalt* class has been identified as more suited to use emerging technologies, like railguns, due to its superior electricity generation capability over previous destroyers and cruisers at 80 megawatts; *Lyndon B. Johnson* specifically was being studied because it is the latest of the class, while the previous two ships would be less likely to initially field the capability due to the testing schedule. The railgun would likely replace one of the two Advanced Gun Systems.[12] By March 2016, construction had become too far along to install the railgun during building, but it can still be added later.[13]

In September 2015, it was reported that Defense of Department officials were considering terminating funding for *Lyndon B. Johnson* prior to its completion.[14] Although considered as a cost-saving measure, cancelling the third *Zumwalt* ship at that stage was likely not possible, and might have ended up actually costing more after paying program shutdown costs and contract termination penalties.[15] By December 2015, the Pentagon had decided in favor of keeping the ship.[16]

3.1.1 References

[1] "Lyndon B Johnson (DDG 1002)". *Naval Vessel Register*. Navy.mil. 12 December 2012. Retrieved 24 November 2014.

[2] Naval Sea Systems Command Office of Corporate Communications (5 April 2012). "Navy Begins Construction on DDG 1002". United States Navy. Retrieved 15 September 2015.

[3] "DDG 1000 Flight I Design". Northrop Grumman Ship Systems. 2007.

[4] Rolls-Royce Marine

[5] "DDG 1001 and DDG 1002 Ship Construction Contract Award Announced" (PDF) (Press release). Naval Sea Systems Command. 15 September 2011. Retrieved 18 April 2012.

[6] "Navy Names Zumwalt Class Destroyer USS Lyndon B. Johnson" (Press release). Defense.gov. 16 April 2012. Retrieved 20 April 2012.

[7] O'Rourke, Ronald (3 February 2012). "Navy DDG-51 and DDG-1000 Destroyer Programs: Background and Issues for Congress" (PDF). *Congressional Research Service*. Federation of American Scientists. p. 42. Retrieved 20 April 2012.

[8] "Work on new destroyer begins". *United Press International* (UPI.com). 11 April 2012. Retrieved 16 April 2012.

[9] "DDG 1000 Zumwalt Class - Multimission Destroyer, United States of America". *Naval-technology.com*. Net Resources International. Retrieved 18 April 2012.

[10] Fabey, Michael (25 January 2013). "U.S. Navy Seeks Alternate Deckhouse For DDG-1002". *Aerospace Daily & Defense Report*.

[11] Cavas, Christopher P. (2 August 2013). "Navy Switches from Composite to Steel". *DefenseNews.com*. Retrieved 15 September 2015.

[12] LaGrone, Sam (5 February 2015). "Navy Considering Railgun for Third Zumwalt Destroyer". *News.USNI.org*. Retrieved 15 September 2015.

[13] Admiral: Shipbuilders won't install railgun on new Navy destroyers - Navytimes, 22 March 2016

[14] Capaccio, Anthony (12 September 2015). "General Dynamics Destroyer Reviewed by U.S. for Cancellation". *Bloomberg News*. Retrieved 15 September 2015.

[15] Cuts To Zumwalt Destroyer Won't Save Much - Breakingdefense.com, 21 September 2015

[16] Pentagon Cuts LCS to 40 Ships, 1 Shipbuilder - Militarytimes.com, 17 December 2015

3.1.2 External links

- U.S. Navy Begins Construction on DDG 1002

- Cutting-edge Navy warship being built in Maine

3.2 USS Michael Monsoor (DDG-1001)

USS *Michael Monsoor* (DDG-1001) is to be the second ship of the *Zumwalt* class of guided missile destroyers. *Michael Monsoor* will be a multi-mission surface combatant tailored for advanced land attack and littoral dominance. The ship's mission is to provide credible, independent forward presence and deterrence and to operate as an integral part of naval, joint or combined maritime forces.

Michael Monsoor will be the second *Zumwalt*-class destroyer. The ship will be 600 feet (180 m) in length, have a beam of 80.7 feet (24.6 m) and displace approximately 15,000 tons. *Michael Monsoor* will have a crew size of 148 officers and sailors; she will make speed in excess of 30 knots (56 km/h; 35 mph).

Michael Monsoor

3.2.1 Etymology

Michael Monsoor is named after Master-at-Arms Second Class Michael A. Monsoor (1981–2006), a United States Navy SEAL killed during the Iraq War and posthumously awarded the Medal of Honor.[4]

3.2.2 Construction

Assembly of modules for *Michael Monsoor* began in March 2010.[5] The keel laying and authentication ceremony for *Michael Monsoor* was held at the General Dynamics-Bath Iron Works shipyard on 23 May 2013.[6] As of May 2013, the *Michael Monsoor* was over 60 percent complete and is scheduled to be delivered in 2016.[6]

3.2.3 References

[1] "NNS130523-14, Future USS Michael Monsoor (DDG 1001) Keel Authenticated". NAVSEA Office of Corporate Communications. 23 May 2013.

[2] "DDG 1000 Flight I Design". Northrop Grumman Ship Systems. 2007.

[3] "RR4500 ship service generator". Rolls-Royce. Retrieved 2014-07-09.

[4] "Michael A. Monsoor". *militarytimes.com*. Military Times. Retrieved 24 December 2010.

[5] "Flash Traffic: Keel Laid for 1st DDG-1000 Destroyer". *The Navy* (Navy Leage of Australia) **74** (1): 15. January 2012. ISSN 1322-6231.

[6] NAVSEA Office of Corporate Communications. "Future USS Michael Monsoor (DDG 1001) Keel Authenticated". Navy.mil. Retrieved 2014-07-09.

- *This article includes information collected from the Naval Vessel Register, which, as a U.S. government publication, is in the public domain. The entry can be found here.*

3.2.4 External links

- NavSource: USS *Michael Monsoor* (DDG-1001)

- USS Monsoor, May 2013

3.3 USS Zumwalt

USS Zumwalt (DDG-1000) is a guided missile destroyer of the United States Navy. She is the lead ship of the *Zumwalt* class and the first ship to be named for Admiral Elmo Zumwalt.[8][9] *Zumwalt* has stealth capabilities, having a radar cross-section akin to a fishing boat despite her large size.[10] On 7 December 2015, *Zumwalt* began its sea trial preparatory to joining the Pacific Fleet.[11] She is to be homeported in San Diego, California.[12]

3.3.1 Etymology

Admiral Elmo Zumwalt

Zumwalt is named after Elmo Russell Zumwalt, Jr., who was an American naval officer and the youngest man to

serve as the Chief of Naval Operations.[13] As an admiral and later the 19th Chief of Naval Operations, Zumwalt played a major role in U.S. military history, especially during the Vietnam War.[13] A highly decorated war veteran, Zumwalt reformed the U.S. Navy's personnel policies in an effort to improve enlisted life and ease racial tensions.[13] After he retired from a 32-year naval career, he launched an unsuccessful campaign for the United States Senate.[13]

The hull classification symbol for *Zumwalt* is DDG-1000, forgoing the guided missile destroyer sequence that goes up to DDG-119 (USS Delbert D. Black, currently the latest of the named *Arleigh Burke*-class destroyers), and continue in the previous "gun destroyer" sequence left off with the last of the *Spruance* class, USS *Hayler*. With the production run of the *Zumwalt* class limited to three units, plans are underway for a third "flight" of *Arleigh-Burke*-class destroyers.

3.3.2 History

See also: Zumwalt-class destroyer § History
Many of the ship's features were originally developed under

Zumwalt's *deckhouse in transit on 6 November 2012*

the DD21 program ("21st Century Destroyer"). In 2001, Congress cut the DD-21 program by half as part of the SC21 program. To save it, the acquisition program was renamed as DD(X) and heavily reworked. The initial funding allocation for DDG-1000 was included in the National Defense Authorization Act of 2007.[14]

The $1.4 billion contract was awarded to Bath Iron Works in Bath, Maine. [15] Full rate production officially began on 11 February 2009.[16]

As of July 2008, the construction timetable was for General Dynamics to deliver the ship in April 2013, with March 2015 as the target for *Zumwalt* to meet her initial operating capability.[17] However, by 2012, the planned completion and delivery of the vessel had slipped to the 2014 fiscal year.[18]

The first section of the ship was laid down on the slipway at Bath Iron Works on 17 November 2011.[18] By this point, fabrication of the ship was over 60% complete.[18] The naming ceremony was planned for 19 October 2013,[19] but was canceled due to the United States federal government shutdown of 2013.[20]

Despite rumors that the launch of *Zumwalt* would be delayed until early 2014, the vessel was launched from its shipyard in Bath, Maine on 29 October 2013.[21][22]

The USS Zumwalt underway for the first time conducting at-sea tests and trials in the Atlantic Ocean Dec. 7, 2015.

In January 2014, *Zumwalt* began to prepare for heavy weather trials. The trials will see how the ship and her instrumentation react to high winds, stormy seas, and adverse weather conditions. The ship's new wave-piercing inverted bow and tumblehome hull configuration reduce her radar cross-section. Tests will involve lateral and vertical accelerations and pitch and roll. Later tests will include fuel onloading, data center tests, propulsion events, X-band radar evaluations, and mission systems activation to finalize integration of electronics, currently 90 percent complete out of 6 million lines of code. These all culminate in builders trials and acceptance trials, with delivery for U.S. Navy tests in late 2014 with initial operating capability (IOC) to be reached by 2016.[23]

Zumwalt's commanding officer is Captain James A. Kirk.[24] Kirk attracted some media attention when he was first named captain, due to the similarity of his name to that of the *Star Trek* television character Captain James T. Kirk, originated by William Shatner. Shatner wrote a letter of support to *Zumwalt*'s crew in April 2014.[25]

On 7 December 2015, the ship departed Bath Iron Works for sea trials to allow the Navy and contractors to operate the vessel under rigorous conditions in order to determine if the Zumwalt is ready to join the fleet as an actively commissioned warship.[11]

On 12 December 2015, during sea trials, *Zumwalt* re-

sponded to a U.S. Coast Guard call for assistance for a fishing boat captain who was experiencing a medical emergency 40 nautical miles (74 km) from Portland, Maine. Due to deck conditions, the Coast Guard helicopter was unable to hoist the patient from the fishing boat, so the *Zumwalt* crew transferred him to the destroyer using an 11-meter rigid-hulled inflatable boat (RHIB), from which he was transported to shore by the Coast Guard helicopter and then to a hospital.[26]

3.3.3 References

[1] "The Navy Just Christened Its Most Futuristic Ship Ever". Business Insider. 2014.

[2] The Navy's New $4.4 Billion Ship Is A Big, Shiny Waste Of Money

[3] Wertheim, Eric (January 2012). "Combat Fleets". *Proceedings* (Annapolis, Maryland: United States Naval Institute) **138** (1): 90. ISSN 0041-798X. Retrieved 13 January 2012. (subscription required (help)).

[4] LaGrone, Sam (21 March 2016). "Zumwalt Destroyer Leaves Bath Iron Works for Builder's Trials". *USNI News*. Retrieved 23 March 2016.

[5] "DDG 1000 Flight I Design". Northrop Grumman Ship Systems. 2007.

[6] Kasper, Joakim (20 September 2015). "About the Zumwalt Destroyer". *AeroWeb*. Retrieved 25 October 2015.

[7] *GAO-05-752R Progress of the DD(X) Destroyer Program*. U.S. Government Accountability Office. 14 June 2005.

[8] "Navy Designates Next-Generation Zumwalt Destroyer". US Department of Defense. 7 April 2006.

[9] "PCU Zumwalt, CAPT James Kirk, Commanding Officer". US Department of Defense. 30 October 2013.

[10] Patterson, Thom; Lendon, Brad (14 June 2014). "Navy's stealth destroyer designed for the video gamer generation". CNN. Retrieved 14 June 2014.

[11] "Largest destroyer built for Navy heads out to sea". *foxnews.com*. Fox News. Retrieved 8 December 2015.

[12] Barber, Elizabeth (30 October 2013). "Navy new destroyer: USS Zumwalt is bigger, badder than any other destroyer". *The Christian Science Monitor*. Retrieved 15 December 2015.

[13] Smith, J.Y. (3 January 2000). "Navy Reformer Elmo Zumwalt Dies". *The Washington Post*. Retrieved 2 October 2014.

[14] NDAA 2007 - "National Defense Authorization Act for Fiscal Year 2007". (109-452) US Government Printing Office. 5 May 2006: 69–70.

[15] "Navy Awards Contracts for Zumwalt Class Destroyers". Navy News Service. 14 February 2008.

[16] "BIW News February 2009" (PDF). General Dynamics Bath Iron Works. 1 March 2009.

[17] "Defense Acquisitions: Cost to Deliver Zumwalt-Class Destroyers Likely to Exceed Budget". Government Accountability Office. 31 July 2008. GAO-08-804

[18] "Flash Traffic: Keel Laid for 1st DDG-1000 Destroyer". *The Navy* (Navy Leage of Australia) **74** (1): 15. January 2012. ISSN 1322-6231.

[19] Cavas, Christopher (3 October 2013). "New Ship News – Sub launched, Carrier prepped, LCS delivered". Defense News.

[20] "Navy Cancels, Postpones Zumwalt Christening". *www. navy.mil*. United States Navy. 11 October 2013. Retrieved 11 October 2013.

[21] "First Zumwalt Class Destroyer Launched". 29 October 2013.

[22] Geoffrey Ingersoll (29 October 2013). "The US Navy's Most Intimidating Creation Yet Just Hit The Water". Business Insider.

[23] DDG 1000 Preps for Heavy Weather Trials - DoDBuzz.com, 14 January 2014

[24] "PCU Zumwalt". US Navy. Retrieved 31 August 2015.

[25] Larter, Davide (16 April 2014). "Famous Capt. Kirk honors real one at ship christening". *Navytimes.com*. Navy Times. Retrieved 8 December 2015.

[26] Miller, Kevin (12 December 2015). "Navy's new Zumwalt rescues ailing fishing boat captain off Portland". Portland Press Herald. Retrieved 13 December 2015.

- *This article includes information collected from the Naval Vessel Register, which, as a U.S. government publication, is in the public domain. The entry can be found here.*

3.3.4   External links

- Official website

- Christening of Lead Ship ZUMWALT (DDG 1000)—Official General Dynamics website

- DDG-1000 Zumwalt / DD(X) Multi-Mission Surface Combatant—GlobalSecurity.org site covering the *Zumwalt* class

- DDG 1000 Zumwalt Class—Multimission Destroyer, United States of America

Chapter 4

Text and image sources, contributors, and licenses

4.1 Text

- **SC-21 (United States)** *Source:* https://en.wikipedia.org/wiki/SC-21_(United_States)?oldid=708566369 *Contributors:* Michael Hardy, Rich Farmbrough, Ketiltrout, David Underdown, SmackBot, Hmains, John, Accurizer, Courcelles, FairuseBot, The ed17, Cydebot, LorenzoB, Toddst1, Dravecky, Ktr101, Dave1185, Fireaxe888, Lightbot, C933103, Yobot, Citation bot 1, Trappist the monk, RjwilmsiBot, Illegitimate Barrister, H3llBot, Palaeozoic99, Helpful Pixie Bot, T-Nod, Metricopolus, Cyberbot II, ChrisGualtieri, Mogism and Anonymous: 9

- **Zumwalt-class destroyer** *Source:* https://en.wikipedia.org/wiki/Zumwalt-class_destroyer?oldid=710289895 *Contributors:* The Epopt, Maury Markowitz, Jinian, Edward, Ixfd64, Mcarling, Davejenk1ns, Samw, PaulinSaudi, Thue, Riddley, Naddy, Xanzzibar, DocWatson42, Akadruid, NightThree, Wwoods, Iceberg3k, Bobblewik, Neilc, Mark5677, The Land, Bbpen, Klemen Kocjancic, Reflex Reaction, O'Dea, Brianhe, Solitude, Rich Farmbrough, ArnoldReinhold, Sarrica, Bender235, Jccooper, Kwamikagami, Kross, TomStar81, Dpaajones, Wiki-Ed, NilsTycho, A2Kafir, Gunter.krebs, Jigen III, LtNOWIS, AN(Ger), Pauli133, Talkie tim, Dan100, Falcorian, Cosal, Fdewaele, Woohookitty, TomTheHand, BlaiseFEgan, GraemeLeggett, BD2412, Nautical, Ketiltrout, Rjwilmsi, Hiberniantears, XLerate, Durin, Mark83, Verybigfish86, OrbitOne, JonathanFreed, Benlisquare, YurikBot, Arado, Epolk, Friedfish, WulfTheSaxon, Gary84, Carajou, Thiseye, Davemck, RL0919, Saberwyn, EEMIV, Lockesdonkey, Svaran, Abune, Kc2hiz, Some guy, Tirronan, Deepdraft, Purpureleaf, SmackBot, Brooksindy, Herostratus, Davidkevin, Cla68, Galloglass, Saros136, Chris the speller, Jprg1966, Thumperward, Snori, AntelopeInSearchOfTruth, Hibernian, Dual Freq, Stephen Hui, Derekbridges, Frap, Glloq, Evil Merlin, Wybot, Henning Makholm, Salamurai, Ohconfucius, STB-1, Accurizer, Scetoaux, Benjaminlobato, PRRfan, Spejic, SandyGeorgia, Intranetusa, Andrwsc, Atakdoug, Tonster, MrDolomite, Zungr, Haus, CP\M, Poweron, Courcelles, Eluchil404, Clay, Byrnejb, HDCase, CmdrObot, Geo8rge, Endeavor51, BeenAroundAWhile, Salmagnone, CompRhetoric, AndrewHowse, Cydebot, Fnlayson, Bob1234321, Hammarbytp, Chad.hutchins, Nottheking, Brad101, Cancun771, Starsword333, Aldis90, The machine512, Kennet.mattfolk, Thijs!bot, Headbomb, Hcobb, Nick Number, EmTeedee, OuroborosCobra, DPdH, Oosh, Dawnseeker2000, Noclevername, Darekun, Wanderingstar04, Akradecki, Mibs, Vamsae, Shirt58, Darklilac, Negator~enwiki, V-man, HolyT, ZZninepluralZalpha, Clairecarpenter, Arch dude, Djkeddie, BilCat, Coldwarrior, DIEXEL, Gwern, MartinBot, Alikaalex, Jogrkim, Ultraviolet scissor flame, Mgibbs, Hifinut, CommonsDelinker, AlexiusHoratius, Mikek999, J.delanoy, Rlsheehan, Mschiavi, NYCRuss, Y2kimmel, Little Miss Might Be Wrong, Notreallydavid, Malthae, Kiwinanday, Signalhead, No1StarWarsfan, Trashbag, ElinorD, Mooremar, Nibios, Thunderbird2, Cvf-ps, Solicitr, Moskevap, Brenont, Sonicology, NoClass, Fanra, KGyST, Lightmouse, TrufflesTheLamb, Reb1981, Smilo Don, Hamiltondaniel, Maralia, Fl3x, Wvfd14, Ossguy, Akeefe98, MBK004, ClueBot, DeaconJohnFairfax, Matrek, Patrick Rogel, Ellivville~enwiki, Guswfla, Jmc41, Ktr101, Eeekster, Iac74205, Simpsonlover2303, Sun Creator, NuclearWarfare, Wirelessenabled, Shem1805, Thingg, Halgin, Emt1299d, DumZiBoT, Coopman86, MystBot, Dave1185, Wiki Mateo, Toyokuni3, Nohomers48, ContiAWB, Blaylockjam10, Colt9033, Lightbot, Pietrow, Zorrobot, Yobot, Akim Dubrow, Mackin90, Nallimbot, Bismarck43, Delta-2030, AnomieBOT, MSP Aviator, Nemesis63, Aneah, ArthurBot, Winged Brick, Nrpf22pr, GrouchoBot, Mark Schierbecker, Lunar Dragoon, Miyagawa, Heroicrelics, Captain Cheeks, Riventree, Wpoihf58, Zerial48, SH9002, Citation bot 1, AstaBOTh15, DrilBot, Foxhound66, Weedle McHairybug, Bmwhd1, Trappist the monk, FFM784, Simtex, RjwilmsiBot, Tvashtar2919, Guikipedia, Superk1a, John of Reading, Rawheas, Pahazzard, Sp33dyphil, TeeTylerToe, ZéroBot, Illegitimate Barrister, Whiteguru, Anir1uph, Krassdaniel, Thewolfchild, Palaeozoic99, FeatherPluma, Ivolocy, ClueBot NG, Brockmvendors, Helpful Pixie Bot, Phd8511, Codepage, Jeancey, 220 of Borg, JonathonSimister, Willard55, America789, Cyberbot II, Adnan bogi, TheJJJunk, Dexbot, Mogism, DickLancaster, NorthBySouthBaranof, The Herald, Pietro13, Fahri Ahmad, UY Scuti, Swiss army fork, Waverapun99, CommanderBill, Monkbot, Eeyoresdream, Llammakey, Aer plene, Gmuwiki55, Lrees4, Mousenight, LocalLaddie, HarryKernow, FORTY5000 and Anonymous: 267

- **Advanced Gun System** *Source:* https://en.wikipedia.org/wiki/Advanced_Gun_System?oldid=708116801 *Contributors:* Riddley, Scott Wilson, Neilc, Discospinster, Rich Farmbrough, ArnoldReinhold, Joshbaumgartner, Ricky81682, Pauli133, Bobrayner, Coemgenus, Mark83, Gurch, Jaraalbe, SmackBot, Emoscopes, Jim62sch, Galloglass, Vechs, Hibernian, Rcbutcher, Fitzhugh, Gbinal, Accurizer, SimonATL, Wikited, Fnlayson, Brad101, Oosh, BilCat, Marcd30319, Notreallydavid, Twfowler, Anooneemiss, The Little Internet Kitty, Lightmouse, Mrniceguy101, William von Zehle, MBK004, Ericvigil, Ost316, Dave1185, Addbot, Mackin90, Srich32977, Mark Schierbecker, Citation bot 1, Tupsumato, Hessamnia, ZéroBot, SporkBot, Mddkpp, UnbiasedVictory and Anonymous: 25

84

- **Bath Iron Works** *Source:* https://en.wikipedia.org/wiki/Bath_Iron_Works?oldid=709119640 *Contributors:* The Epopt, Jinian, Michael Hardy, Notheruser, Robbot, Decumanus, NightThree, Wwoods, Iceberg3k, Bbpen, Donan.raven, D6, N328KF, Rich Farmbrough, Rje, PaulHanson, Joshbaumgartner, Gulfstorm75, Pjmorse, Kurieeto, Woohookitty, Bellhalla, Tim!, Valentinejoesmith, Boatman, XLerate, MLRoach, Cyber-prog, Epipelagic, Albyva, SmackBot, Finavon, Gjs238, Florian Adler, Famspear, Derekbridges, Rhollenton, Jwillbur, Ligulembot, Publicus, PRRfan, Neddyseagoon, Haus, Octane, Namiba, HennessyC, ShelfSkewed, Thijs!bot, Sizuru~enwiki, Magioladitis, Flowanda, R'n'B, Wil101, Chiswick Chap, Nono le petit robot~enwiki, A4bot, SQL, SE7, Lightmouse, Maralia, Xnatedawgx, ImageRemovalBot, MBK004, Plastikspork, Masterblooregard, Ktr101, Excirial, Thewellman, Romney yw, Life of Riley, Julienman11, Cmr08, Addbot, Lightbot, Shannon1, Yobot, Ptbotgourou, Николай 98765, Citation bot, Mark Schierbecker, Full-date unlinking bot, Cnwilliams, Lotje, ZéroBot, $1LENCE D00600D, Palaeozoic99, Sven Manguard, Sharkmouth, Helpful Pixie Bot, Calidum, BG19bot, CitationCleanerBot, WestportWiki, Hmainsbot1, Skimaine15, Peteramainer and Anonymous: 31

- **Elmo Zumwalt** *Source:* https://en.wikipedia.org/wiki/Elmo_Zumwalt?oldid=708811339 *Contributors:* Jeronimo, Imran, Jinian, Hephaestos, Docu, Jiang, Loren Rosen, Matithyahu, Huangdi, Donreed, Flauto Dolce, Wally, Davidcannon, Nunh-huh, Folks at 137, Mboverload, PDH, Mark5677, Rlquall, Husnock, Necrothesp, Neutrality, Marine 69-71, Klemen Kocjancic, Mtnerd, D6, Rich Farmbrough, Björn Knutson, Tom, Lokifer, PaulHanson, Ricky81682, Andrew Gray, TommyBoy, Pmeisel, ItemCo16527, Suruena, Inge, EECEE, Neanderthalprimadonna, GraemeLeggett, Dysepsion, Ted Wilkes, Miq, Lockley, Tony619, Ground Zero, NekoDaemon, Whateley23, Robert Prummel, RobyWayne, MikeyChalupa, Stevenfruitsmaak, RussBot, Arado, DanMS, Howcheng, Breathstealer, Gadget850, E-Dogg, Rms125a@hotmail.com, Thomas Blomberg, John Broughton, MatthewSMaynard, SmackBot, Looper5920, GrummelJS, KocjoBot~enwiki, AustinKnight, Kintetsubuffalo, Jkp1187, ERcheck, Schmiteye, Rlevse, John Reaves, Jwillbur, Severinus, Ala.foum, Xdamr, Jamestown, Nobunaga24, PRRfan, Spiff666, BrownCow, Djharrity, Nehrams2020, Mcwatson, CmdrObot, Drinibot, Mhenneberry, Chicheley, Cydebot, George Al-Shami, Hebrides, Lugnuts, Hometack, Brad101, Alexander lau, RobotG, Xhienne, Joebengo, Buckshot06, Waacstats, Markus Becker02, GhostofSuperslum, PMG, CommonsDelinker, Marcd30319, HenryLarsen, Thismightbezach, Magnet For Knowledge, EricSerge, MosesHall, Mimich, Monsieurdl, Eguler, Bennypea, Gbawden, SE7, Toddst1, Monegasque, Claudevsq, Kumioko (renamed), Aumnamahashiva, LarRan, MBK004, Cuprum17, Matrek, All Hallow's Wraith, AusTerrapin, Drmies, Gnome de plume, SpikeToronto, Sun Creator, M.O.X, Searcher 1990, Thewellman, 1ForTheMoney, Wjwtk, StudNamedAaron, Good Olfactory, Kbdankbot, Addbot, AntonyZ, Lightbot, Yobot, Fraggle81, Dodgerblue777, Donfbreed, AnomieBOT, Femmefabuleuse, Валерий Пасько, LilHelpa, FreeRangeFrog, Obersachsebot, Jayarathina, 219.106??, Armbrust, Jamesh15777, Mikie yorkie, Bheuninckx, Full-date unlinking bot, DocYako, Canuckian89, RjwilmsiBot, Acsian88, Pearl Dragon, EmausBot, Illegitimate Barrister, Bullmoosebell, Δ, Sven Manguard, AusTerrapinBotEdits, ClueBot NG, TucsonDavid, ANGELUS, Asalrifai, McOleo, Zeraful, JakeInJoisey, ChrisGualtieri, Mpollow, Mogism, Faizan, ArmbrustBot, A0852741963, BrettASnyder, KasparBot and Anonymous: 98

- **Free-electron laser** *Source:* https://en.wikipedia.org/wiki/Free-electron_laser?oldid=712015415 *Contributors:* Bryan Derksen, Julesd, Reddi, Hankwang, Academic Challenger, GreatWhiteNortherner, Alan Liefting, DocWatson42, Jpatej, Leonard G., Micru, Peter bertok, Deglr6328, Deleteme42, Pjacobi, Xezbeth, Kghose, DaveGorman, Kjkolb, Pearle, Hooperbloob, Gene Nygaard, Tylerni7, J M Rice, Graham87, Tommcnabb, Rjwilmsi, China Crisis, KaiMartin, Arnero, Nwatson, Lmatt, David H Braun (1964), Meawoppl, DVdm, Roboto de Ajvol, Shaddack, Eleassar, David R. Ingham, Welsh, RabidDeity, Santaduck, Engineer Bob, Erik J, Saikiri, SmackBot, Mjspe1, Gregjgrose, Kmarinas86, Oatmeal batman, Hgrosser, Hawkwings31, DMacks, Giancarlo Rossi, Tomatoman, JorisvS, Hu12, Chetvorno, CmdrObot, Thijs!bot, Headbomb, Dtgriscom, Second Quantization, Noclevername, Guy Macon, Lfstevens, Albertvillanovadelmoral, Freddy011, Schmloof, Pagw, Alexander Patrakov, R'n'B, Jon-e-five, Mjgullans, Stuffysour, Lantonov, DadaNeem, VolkovBot, Landisdesign, TXiKiBoT, Enozkan, Hqb, Revansx~enwiki, Jerryobject, Lightmouse, KJG2007, Polyamorph, Alexbot, Gabella, Davismargaret, XLinkBot, NellieBly, Dyuku, Bonewith, Luckas-bot, AnomieBOT, Obersachsebot, Xqbot, Tomdo08, J04n, Trurle, Xfig, Pozharnikar, Coosbane, Kyteto, Citation bot 1, Pinethicket, Jonesey95, JMMuller, Michael9422, RjwilmsiBot, Paratwa, EmausBot, John of Reading, Karim osama1, WikitanvirBot, ZéroBot, Wikfr, RaptureBot, Wakebrdkid, BR84, ClueBot NG, Powersjcb, ArbHH, Marieto60, MrBill3, Jannick88, Comfr, Dexbot, Ziggy1986 TS, Tony Mach, GravRidr, FizykLJF, Maderthaner, Seneszenz, Ggf4t and Anonymous: 99

- **Guided missile destroyer** *Source:* https://en.wikipedia.org/wiki/Guided_missile_destroyer?oldid=711603061 *Contributors:* Jinian, Minesweeper, Pcb21, Stan Shebs, Angela, Conti, TomD1939, StinKerr, MK~enwiki, Seglea, Niteowlneils, Bbpen, D6, Warpflyght, Kross, TomStar81, MarkGallagher, Choess, Chwyatt, Borgx, Hairy Dude, Arado, Ksyrie, Spot87, BCGarvey, Nekonobaka, Windyjarhead, Andrewkantor, By78, Bluebot, Thom2002, Jprg1966, Enomosiki, Dual Freq, OrphanBot, Mtmelendez, Accurizer, Beetstra, Courcelles, Blicious, Xnuiem, HolyT, Chanakyathegreat, BilCat, Bot-Schafter, KylieTastic, Ejh0819~enwiki, Usergreatpower, Solicitr, Signsolid, Hockeyroger, WereSpielChequers, Benea, Reb1981, 61mei31, MBK004, Sturmvogel 66, Shem1805, Callinus, Addbot, Stephen Fulcher, LatitudeBot, Enthusiast10, Mackin90, Kadrun, Ulric1313, LilHelpa, Chen Guangming, Mr George R. Allison, Skcpublic, FrescoBot, Ironboy11, Bambuway, OgreBot, Bcs09, DexDor, Illegitimate Barrister, Space25689, AvicAWB, Quite vivid blur, $1LENCE D00600D, Strike Eagle, BG19bot, Mich.kramer, Dexbot, Ratyuihgf, Sudhanshu Nimbalkar, Sam Sailor, UY Scuti, Mularam2014, Filedelinkerbot, Fried leek, Llammakey, VandeMataram, Craftwerker, Peter O'Conner, ????, Nani1992 and Anonymous: 74

- **Huntington Ingalls Industries** *Source:* https://en.wikipedia.org/wiki/Huntington_Ingalls_Industries?oldid=708601992 *Contributors:* Smithd, Wwoods, Anthony Appleyard, TommyBoy, Kenyon, Tim!, XLerate, Arado, Teb728, Quidam65, Gump Stump, Woodshed, Danrok, Conquistador2k6, Brad101, Hcobb, BilCat, Shortride, TXiKiBoT, Timhogs, Avhell, Matrek, Niceguyedc, Sun Creator, Svgalbertian, Yobot, Ptbotgourou, Jean.julius, AnomieBOT, RobinInTexas, AdmiralHood, Dgat16, Skcpublic, Bstaghare, Mean as custard, Djembayz, Anir1uph, Rangoon11, ClueBot NG, Loginnigol, Sabre ball, BG19bot, Njacobs1, BattyBot, Tow, Dexbot, TBSorensen, PghPhxNfk, Fashionsforward, RichyForecast and Anonymous: 13

- **Nunn–McCurdy Amendment** *Source:* https://en.wikipedia.org/wiki/Nunn%E2%80%93McCurdy_Amendment?oldid=694780912 *Contributors:* Jeffq, Thorwald, Stephan Leeds, Fivemack, Mais oui!, Davewild, PRRfan, Alaibot, MCG, I, Podius, MNeimeyer, RP459, MystBot, Addbot, Luckas-bot, Yobot, Cantons-de-l'Est, FrescoBot, Koakhtzvigad, ChrisGualtieri, JackBrad419, ArmbrustBot and Anonymous: 6

- **Railgun** *Source:* https://en.wikipedia.org/wiki/Railgun?oldid=710688195 *Contributors:* Bryan Derksen, JeLuF, SimonP, DrBob, Heron, Chris Q, Frecklefoot, Patrick, JohnOwens, Gabbe, Ixfd64, Baylink, Erzengel, Julesd, Glenn, Tristanb, GCarty, Erzengel, Elvis, Wikiborg, Fuzheado, Zoicon5, Maximus Rex, Ed g2s, Thue, Nickshanks, AnthonyQBachler, Jeffq, Shantavira, Vt-aoe, Fredrik, Chris 73, Nurg, Lowellian, Postdlf, Spike, Premeditated Chaos, Bertie, Hadal, Cyrius, Carnildo, Ancheta Wis, Johnjosephbachir, DocWatson42, YanA, Wolfkeeper, Tom harrison, Leonard G., Gracefool, Foobar, Bobblewik, Neilc, Physicist, Roisterer, Gem fr, SAMAS, Darksun, Joyous!, Sonett72, Grm wnr, SYSS Mouse, Zowie, NathanHurst, CannedLizard, Rich Farmbrough, Qutezuce, Vsmith, Wikiacc, Fluzwup, Alistair1978, Bender235, Neko-chan, Sockatume,

Solidus~enwiki, Kross, Bookofjude, Thunderbrand, Solra Bizna, TomStar81, Slicky, Kjkolb, Martyman, Glaucus, Anthony Appleyard, Redxiv, Geo Swan, Joshbaumgartner, Axl, Hinotori, Ferrierd, Mac Davis, Hu, Hohum, NAshbery, Danntm, Cal 1234, RJFJR, Pauli133, DV8 2XL, Gene Nygaard, Dziban303, Capecodeph, Axeman89, Bew, Crosbiesmith, Cl~enwiki, Bobrayner, OwenX, Mikeetc, Nipsy, Dandv, Benbest, Jeff3000, Insomniac By Choice, GregorB, Scm83x, Arrkhal, M412k, GraemeLeggett, Emerson7, Royan, Ashmoo, Kbdank71, Ratamacue, De-Piep, Sjakkalle, Rjwilmsi, Koavf, Bruce1ee, Ligulem, MarnetteD, JYOuyang, Mark83, Ewlyahoocom, Sdr, Zotel, WouterBot, Chobot, Bgwhite, Peterl, YurikBot, RussBot, DMahalko, Arado, Killervogel5, RealMontrealer, RadioFan, Gaius Cornelius, GunnarRene, Capi, Daemon8666, Dudtz, Cleared as filed, Irishguy, Moe Epsilon, Saberwyn, Salmanazar, NorsemanII, Le Blue Dude, Light current, Arthur Rubin, GraemeL, JoanneB, CWenger, Mozkill, JLaTondre, Junglecat, Eigenlambda, Dr Napalm, SmackBot, Prodego, TestPilot, Pgk, Angelstorm, Lengis, Dark-lock, Arny, Athaler, Galloglass, Gilliam, Squiddy, Marc Kupper, Schmiteye, Chris the speller, Achmelvic, Thumperward, Hibernian, Karmon, George Church, Jerome Charles Potts, Oni Ookami Alfador, Dual Freq, Emurphy42, Can't sleep, clown will eat me, Egsan Bacon, Crad0010, Chlewbot, Onorem, Nakon, PointyOintment, RocketGuy, A5b, Shawn2082, Daniel.Cardenas, J 1982, Trassiorf, Kashmiri, Shadowspirit216, Jehar, Rock4arolla, Tonster, Krispos42, MaverickPH, JYi, Kernow, Guglido, Andrew Hampe, Courcelles, Pathosbot, Tawkerbot2, Deon, Schmil~enwiki, IntrigueBlue, N2e, Old Guard, Dan Fuhry, Malamockq, Fnlayson, CovenantD, Daniel J. Leivick, Benvogel, Gnfnrf, Thijs!bot, Kubanczyk, Archmichael, Jmg38, General Deathstroke, MaggoT SeveN, Gamer007, Nonagonal Spider, Tellyaddict, Hcobb, Alexezeoke, Nick Number, Trakon, Stannered, Fireplace, Luna Santin, Guy Macon, Marokwitz, Railgunner, Mr Grim Reaper, Darklilac, Dodonpachi, Lklundin, Ingolfson, Res2216firestar, Omeganian, Shardakar, Dream Focus, Magioladitis, VoABot II, BigDukeSix, Boomcoach, Yandman, Kyanwan, Fa-ther Goose, NervousSystem, Mtd2006, Gphoto, Macmelvino, MartinBot, Buddhahat, Drew1369, Superstuwy, 🔲🔲🔲🔲, Jim.henderson, Sm8900, Francis Tyers, Uriel81~enwiki, Thoth171, Odichthys, Slash, J.delanoy, Nev1, Acalamari, Tarotcards, Mrg3105, AntiSpamBot, Wariner, Inspec-torTiger, Nwbeeson, Mjrchaos, Psminson, Vanished user 39948282, Cinnagingercat, BenFenner, RjCan, Halmstad, Devgil, Caribbean H.Q., VolkovBot, Derekbd, Philip Trueman, Timventura, RagnarokEOTW, Ann Stouter, Pah246, LeaveSleaves, Natg 19, Wingedsubmariner, Lite-FireDark, Daveletourneau~enwiki, RiverStyx23, Buffs, HaPpY-ToWn, SQL, Seekedanddestroyed, Squalk25, Logan, Cunjo, SieBot, Sonicology, Scarian, Unregistered.coward, Yerpo, Superllama, Scorpion451, Avnjay, Nuttycoconut, Hamiltondaniel, Treekids, Denisarona, Pat1717, Clue-Bot, Matrek, Nitack, Wanderer57, Chessy999, Mild Bill Hiccup, Gerbilo, LizardJr8, Ericvigil, Anonymous101, Robbie098, Bertillini, The Founders Intent, NuclearWarfare, StanContributor, Staygyro, Lukipuk, Carriearchdale, Footballfan190, Deathmare, Nuketime201, Freelion, DumZiBoT, Dileepvr, XLinkBot, Ost316, Fresnel149, Facts707, SilvonenBot, Big Ed Wiki, Jasper2101, Addbot, Mr0t1633, Chicago3141, Melab-1, Superduperpup, Aloha2436, Bwrs, Lightbot, ScienceApe, Genius101, Hyperdimensionalentity, Yobot, Mackin90, EnochBethany, JBancroftBrown, Railgunchanger, Cock1111, AnomieBOT, DemocraticLuntz, Archon 2488, Uikku, Jim1138, Piano non troppo, Asgard10, Slimey662, Materialscientist, Carlsotr, Eumolpo, I ate a glabnorgog, Xqbot, Bihco, Irishguy117, Alkibiades231, Frosted14, Mark Schierbecker, Erik9, Wolfman93, Tobias schoepfer, GliderMaven, FrescoBot, LucienBOT, 🔲🔲🔲, ProtoDrake, Citation bot 1, Tom.Reding, Roboo.jack, PentagonParanormal, Michael J. Chapman, Full-date unlinking bot, Kielbasa1, Reaper Eternal, ThinkEnemies, TGCP, Acather96, Wikitanvir-Bot, Tobias559, Rolir, RA0808, Klbrain, Mark Sitton, Arkenflame, Shuipzv3, Yitzachmeyer, Ὁ οἶστρος, Zarlachan, SporkBot, Wayne Slam, Wingman417, Emptyecho, Efenna, Orange Suede Sofa, Rangoon11, ChuispastonBot, Teapeat, Whoop whoop pull up, ClueBot NG, Xplor-rCur8, Michaelmas1957, El Roih, Shaddim, Frietjes, MTN1996, ElectroWolf, 101japple, Ose\fio, Danim, Helpful Pixie Bot, Martin Berka, Calabe1992, Dcbacker, BG19bot, Virtualerian, Gsimpso4, SunTzuArtOfWar, Cadiomals, TROPtastic, Zedshort, Conifer, Aisteco, Fylbecatu-lous, Spock1995, BattyBot, America789, Cyberbot II, ChrisGualtieri, Sundibar, Thegamer117, Mrphious, Deathcat101, Werewolf Astronaut, Joshtaco, Frosty, Knobeeoldben, Smool-37, Harmanjot singh kalra, Abishai 300, AmaryllisGardener, Kylethekillman, Kahtar, Tazzik Salarian, Cousteau, Agent0047, Yossarian-4096, The Oaked Ridge, Robb.main, HelenRMcLoud, Nuclear12321, Donosauro, Mario Castelán Castro, Uzair3254, Babarstamp52, Msfredenburg, Tralala0, CLCStudent, Tfocker4, Jhallard, Trollo Baggins and Anonymous: 680

- **Stealth ship** *Source:* https://en.wikipedia.org/wiki/Stealth_ship?oldid=698068671 *Contributors:* Denni, Warmfuzzygrrl, Nurg, Andries, Laisak, Filur, Nrbelex, Fawcett5, Talkie tim, Sylvain Mielot, Rjwilmsi, Mark83, YurikBot, Arado, BeBop~enwiki, Splash, Kurt Leyman, Kirill Lokshin, BOT-Superzerocool, Caerwine, Dahlis, Orcaborealis, Mossig, Alureiter, SmackBot, Commander Keane bot, Amatulic, OrphanBot, MJBurrage, Andrew c, Shrew, Joffeloff, Haus, CP\M, Courcelles, Henrickson, Oden, Cydebot, Aldis90, EmTeedee, Ingolfson, Chanakyathegreat, Db099221, Jacce, Indon, BilCat, Pikolas, CommonsDelinker, KTo288, J.delanoy, KylieTastic, Xelous, Khutuck, Oom Agent, Meatwaggon, DirkV, Dpope2, Maralia, ImageRemovalBot, Fredmdbud, ClueBot, PipepBot, Shem1805, BOTarate, PauliKL, DumZiBoT, WikHead, Dave1185, Addbot, Tide rolls, Luckas Blade, Aday, The Bushranger, Yobot, عالم محبوب, AnomieBOT, Theoprakt, Pipeafcr, ArthurBot, Jeffrey Mall, RibotBOT, FrescoBot, D'ohBot, OgreBot, Alonso de Mendoza, LittleWink, Latios, Goodboy2009, சஞ்தி, Jakerin, EmausBot, WikitanvirBot, UltimaRatio, Leechangung, Gsarwa, ChuispastonBot, Strike Eagle, Harizotoh9, Klilidiplomus, Correctiondetail, Smithsfamousfarm, Cyberbot II, JimnChina, Waverapun99, Filedelinkerbot, Sciophobiaranger, BobNesh and Anonymous: 76

- **Tumblehome** *Source:* https://en.wikipedia.org/wiki/Tumblehome?oldid=711018554 *Contributors:* Maury Markowitz, Timc, Greglocock, Xanzzibar, DocWatson42, Iceberg3k, O'Dea, Brianhe, Rama, Kappa, Msylvester, Pauli133, Falcorian, Woohookitty, Nautical, Dar-Ape, RJP, Arado, Icarus3, Hellbus, DAJF, DeadEyeArrow, Petri Krohn, Tevildo, Groyolo, SmackBot, Herostratus, Eaglizard, Peter Isotalo, Bluebot, Thumperward, Rcbutcher, Derekbridges, JustinSmith, Woodshed, Cydebot, Aldis90, Hcobb, Joel Bradshaw, Widefox, Corella, LorenzoB, Mar-tocticvs, Rlsheehan, GS3, Nibios, SE16, Fenwayguy, Ktr101, Addbot, Luckas-bot, Moyasta, QueenCake, Tongshan, Latios, Micraboy, Midnight bird, BP OMowe, Ri Osraige, Roguenode and Anonymous: 31

- **United States naval gunfire support debate** *Source:* https://en.wikipedia.org/wiki/United_States_naval_gunfire_support_debate?oldid= 707799426 *Contributors:* Delirium, PaulinSaudi, Tpbradbury, Michael Devore, Hammersfan, Rich Farmbrough, TomStar81, Ashley Pomeroy, Woohookitty, Kelisi, Mandarax, Rjwilmsi, Bgwhite, Tony1, Sperril, Petri Krohn, SmackBot, Melchoir, Cla68, Chris the speller, Hibernian, Dual Freq, Derekbridges, Robofish, Neovu79, SandyGeorgia, BranStark, Adam sk, CmdrObot, The ed17, Cydebot, AtTheAbyss, Epbr123, Hcobb, HolyT, Kaleja, Giggy, David Eppstein, R'n'B, KTo288, Patar knight, Johnbod, FergusM1970, Toddy1, Triesault, Bahamut0013, Brenont, Ravens-fire, Lightmouse, Dabomb87, MBK004, Niceguyedc, Cirt, Ktr101, Lightbot, Yobot, Mackin90, AnomieBOT, KenLee318, Citation bot, Mr. Military, RightCowLeftCoast, Citation bot 1, John of Reading, H3llBot, Helpful Pixie Bot, BattyBot, Cyberbot II, ChrisGualtieri, Monkbot and Anonymous: 28

- **Vertical launching system** *Source:* https://en.wikipedia.org/wiki/Vertical_launching_system?oldid=707327847 *Contributors:* Patrick, Sky-smith, DocWatson42, Everyking, Iceberg3k, Tagishsimon, RobinCarmody, Balcer, Imjustmatthew, Mtnerd, Quilche, N328KF, Discospin-ster, Avriette, Rama, Moki80, ArnoldReinhold, Bender235, Sum0, Chairboy, TomStar81, Sasquatch, Kjkolb, Kitplane01, C-4, Hooperbloob, A2Kafir, Wendell, Grutness, Kazuaki Shimazaki, Ashley Pomeroy, JanSöderback, Vedant, LordAmeth, Admiral Valdemar, Firsfron, Poc-cilScript, Rjwilmsi, Drbrain, StuartBrady, Demarchist, Mark83, Arado, John Smith's, Opheleum, FrisoHoltkamp, Hellbus, Hydrargyrum, Ksyrie,

Ergbert, Saberwyn, Datafuser, Ageekgal, Pyeknu, Shyam, Nick-D, Tom Morris, SmackBot, Emoscopes, Deon Steyn, Mdd4696, By78, Grant Janzen, Moshe Constantine Hassan Al-Silverburg, Scienz Guy, Dual Freq, OrphanBot, RedHillian, Earthworm Makarov, Accurizer, Octane, Maxrandom777, Damifb, Underpants, Cancun771, Aldis90, Kirk Hilliard, Thijs!bot, SGGH, Dickhooker, Escarbot, Mongreldog, Kennethij, Chanakyathegreat, Gsking, Two way time, BilCat, Frotz, Subspace1250, Afterthewar~enwiki, Gwern, S3000, Walle83, CommonsDelinker, Marcd30319, Idioma-bot, Imperator3733, Usertaffy3, Broadbot, Heb, RucasHost, ImageRemovalBot, Bskipper, MBK004, Masterblooregard, Ktr101, Taifarious1, Res Gestæ Divi Augusti, Jmkim dot com, MystBot, Addbot, JBsupreme, Greyhood, Yobot, امارات|1971, Bismarck43, Arjun G. Menon, Sz-iwbot, Blitzoace, High Contrast, AdmiralHood, Johnxxx9, GrouchoBot, Parabellum101, LucienBOT, Trappist the monk, John of Reading, Look2See1, Putaro, 􏿿􏿿􏿿, Vasky22, Krassdaniel, Zhaguix, ClueBot NG, Strike Eagle, IridiumIs77, BG19bot, Pine, Hsasar, Adnan bogi, DelamontagneNL, Randolph69, 􏿿􏿿􏿿, Nicky mathew, Editor abcdef, Hatchiko, Peter O'Conner and Anonymous: 122

- **USS Lyndon B. Johnson (DDG-1002)** *Source:* https://en.wikipedia.org/wiki/USS_Lyndon_B._Johnson_(DDG-1002)?oldid=711444424 *Contributors:* Mcarling, Wwoods, Benlisquare, MrDolomite, Cydebot, Brad101, Hcobb, Oosh, Grishnackh, Parsecboy, RP88, Addbot, Yobot, Thehelpfulbot, Trappist the monk, Sp33dyphil, Illegitimate Barrister, America789, Cyberbot II, TeriEmbrey, Monkbot, Llammakey, Aer plene, Luis Santos24 and Anonymous: 4

- **USS Michael Monsoor (DDG-1001)** *Source:* https://en.wikipedia.org/wiki/USS_Michael_Monsoor_(DDG-1001)?oldid=699220919 *Contributors:* Mcarling, Wwoods, TiMike, Benlisquare, Saberwyn, Pyeknu, MrDolomite, Courcelles, Eastlaw, Location, Cydebot, Oosh, Parsecboy, MBK004, Fltadm, Download, Lightbot, Mackin90, Delta-2030, AnomieBOT, Safiel, Captain Cheeks, DrilBot, Trappist the monk, Ekanzoy, Illegitimate Barrister, SporkBot, Palaeozoic99, BattyBot, America789, TeriEmbrey, Monkbot, Llammakey, Luis Santos24 and Anonymous: 13

- **USS Zumwalt** *Source:* https://en.wikipedia.org/wiki/USS_Zumwalt?oldid=711977043 *Contributors:* Jinian, Mcarling, DocWatson42, Wwoods, Macrakis, Iceberg3k, Bigpeteb, Kiteinthewind, Vsmith, Adornix, TomStar81, Raymond, Skyring, Falcorian, Cosal, Bellhalla, Nautical, Rjwilmsi, Benlisquare, Zimbabweed, Sendell~enwiki, Davemck, Saberwyn, Arthur Rubin, Streltzer, Havocrazy, Davidkevin, Cla68, Thumperward, Martin Blank, Ohconfucius, John, Accurizer, Scetoaux, PRRfan, Intranetusa, Tonster, MrDolomite, Haus, Courcelles, Hornblende, Cydebot, Chad.hutchins, Brad101, Epbr123, Headbomb, Hcobb, Oosh, Widefox, HolyT, Nthep, TAnthony, Parsecboy, RP88, Psycardis, Subman758, Kimse, Marcd30319, Navy1775, Jvcdude, Bart-16, Andy Marchbanks, Hammersoft, ElinorD, SGT141, SwordSmurf, AlleborgoBot, Moskevap, Lightmouse, TrufflesTheLamb, Mrniceguy101, Reb1981, Maralia, FlamingSilmaril, Scillystuff, MBK004, De728631, PipepBot, TypoBoy, Ktr101, Another Believer, Addbot, Nohomers48, Reedmalloy, Download, Farmercarlos, Colt9033, Lightbot, Mackin90, National security geek, Penguino35, Delta-2030, AnomieBOT, Николай 98765, Cdevers, JTBowen, Captain Cheeks, Citation bot 1, DrilBot, Johnny2zk, Trappist the monk, RjwilmsiBot, TeeTylerToe, Illegitimate Barrister, Thewolfchild, Palaeozoic99, ClueBot NG, TMX-Mike, Dnewell78, Calidum, DSkauai, JonathonSimister, America789, Hayward62, Cyberbot II, Joedumlao, Adnan bogi, Padenton, Khazar2, Veronicawilson235, Kelvin5452, Monkbot, SantiLak, Sciophobiaranger, Llammakey, Jannidd, FPS James Bond 007, Joemariner1984, Troywsimpson and Anonymous: 82

4.2 Images

- **File:120416-N-AL577-001.jpg** *Source:* https://upload.wikimedia.org/wikipedia/commons/9/94/120416-N-AL577-001.jpg *License:* Public domain *Contributors:* http://www.navy.mil/view_image.asp?id=121807 *Original artist:* Lt. Shawn Eklund

- **File:3d-radarp.jpg** *Source:* https://upload.wikimedia.org/wikipedia/commons/1/11/3d-radarp.jpg *License:* CC-BY-SA-3.0 *Contributors:* Transferred from de.wikipedia to Commons by Wdwd using CommonsHelper. *Original artist:* Christian Wolff

- **File:5-54-Mark-45-firing_edit.jpg** *Source:* https://upload.wikimedia.org/wikipedia/commons/8/8e/5-54-Mark-45-firing_edit.jpg *License:* Public domain *Contributors:*
This Image was released by the United States Navy with the ID 970418-N-4142G-002 <a class='external text' href='//commons.wikimedia.org/ w/index.php?title=Category:Files_created_by_the_United_States_Navy_with_known_IDs,<span>,&,</span>,filefrom=970418-N-4142G-002#mw-category-media'>(next)</a>.
This tag does not indicate the copyright status of the attached work. A normal copyright tag is still required. See Commons:Licensing for more information.
Original artist: U.S. Navy, Photographer's Mate 2nd Class Felix Garza Jr

- **File:A_device.svg** *Source:* https://upload.wikimedia.org/wikipedia/commons/e/e7/A_device.svg *License:* Public domain *Contributors:* (1953 edition with updates through 1960) *NavPers 15,790: Navy and Marine Corps Awards Manual* (PDF), Washington, DC: Department of the Navy, pp. 57–59 Retrieved on 15 May 2009. OCLC: 45726498. *Original artist:* Oneam

- **File:Advanced_Electric_Ship_Demonstrator.jpg** *Source:* https://upload.wikimedia.org/wikipedia/commons/4/4a/Advanced_Electric_Ship_Demonstrator.jpg *License:* Public domain *Contributors:* Navy NewsStand Photo ID: 051130-N-7676W-081
Navy NewsStand Home *Original artist:* United States Navy, John F. Williams

- **File:Ambox_current_red.svg** *Source:* https://upload.wikimedia.org/wikipedia/commons/9/98/Ambox_current_red.svg *License:* CC0 *Contributors:* self-made, inspired by Gnome globe current event.svg, using Information icon3.svg and Earth clip art.svg *Original artist:* Vipersnake151, penubag, Tkgd2007 (clock)

- **File:Ambox_important.svg** *Source:* https://upload.wikimedia.org/wikipedia/commons/b/b4/Ambox_important.svg *License:* Public domain *Contributors:* Own work, based off of Image:Ambox scales.svg *Original artist:* Dsmurat (talk · contribs)

- **File:American_Defense_Service_ribbon.svg** *Source:* https://upload.wikimedia.org/wikipedia/commons/8/8e/American_Defense_Service_ribbon.svg *License:* Public domain *Contributors:* Vectorized from raster image <a href='//commons.wikimedia.org/wiki/File:USADefRib.gif' class='image'><img alt='USADefRib.gif' src='https://upload.wikimedia.org/wikipedia/commons/5/54/USADefRib.gif' width='106' height='30' data-file-width='106' data-file-height='30' /></a> *Original artist:* Ipankonin

- **File:Asiatic-Pacific_Campaign_ribbon.svg** *Source:* https://upload.wikimedia.org/wikipedia/commons/0/08/Asiatic-Pacific_Campaign_ribbon.svg *License:* Public domain *Contributors:* Vectorized from raster image <a href='//commons.wikimedia.org/wiki/File:AsiaCampRib.png' class='image'><img alt='AsiaCampRib.png' src='https://upload.wikimedia.org/wikipedia/commons/thumb/3/39/AsiaCampRib.png/

96px-AsiaCampRib.png' width='96' height='27' srcset='https://upload.wikimedia.org/wikipedia/commons/3/39/AsiaCampRib.png 1.5x, https://upload.wikimedia.org/wikipedia/commons/3/39/AsiaCampRib.png 2x' data-file-width='106' data-file-height='30' /></a> AsiaCampRib.gif *Original artist:* Ipankonin

- **File:Asrocnuke1962.jpg** *Source:* https://upload.wikimedia.org/wikipedia/commons/3/37/Asrocnuke1962.jpg *License:* Public domain *Contributors:* Transferred from en.wikipedia to Commons. *Original artist:* The original uploader was Tempshill at English Wikipedia

- **File:Award_star_(gold).png** *Source:* https://upload.wikimedia.org/wikipedia/commons/d/d5/Award_star_%28gold%29.png *License:* CC BY-SA 3.0 *Contributors:* This file was derived from: Award star-gold-3d.svg
  Original artist: This image includes elements that have been taken or adapted from this: <a href='//commons.wikimedia.org/wiki/File:Award_star-gold-3d.svg' class='image'><img alt='Award star-gold-3d.svg' src='https://upload.wikimedia.org/wikipedia/commons/thumb/a/a2/Award_star-gold-3d.svg/21px-Award_star-gold-3d.svg.png' width='21' height='20' srcset='https://upload.wikimedia.org/wikipedia/commons/thumb/a/a2/Award_star-gold-3d.svg/31px-Award_star-gold-3d.svg.png 1.5x, https://upload.wikimedia.org/wikipedia/commons/thumb/a/a2/Award_star-gold-3d.svg/42px-Award_star-gold-3d.svg.png 2x' data-file-width='535' data-file-height='510' /></a> Award star-gold-3d.svg.

- **File:Biw_aerial.jpg** *Source:* https://upload.wikimedia.org/wikipedia/commons/b/b7/Biw_aerial.jpg *License:* Public domain *Contributors:* U.S. Navy Naval Air Station Brunswick website [1] *Original artist:* USN

- **File:Bronze-service-star-3d.png** *Source:* https://upload.wikimedia.org/wikipedia/commons/b/b0/Bronze-service-star-3d.png *License:* CC BY-SA 3.0 *Contributors:* Own work *Original artist:* Lestatdelc

- **File:Bronze_Star_ribbon.svg** *Source:* https://upload.wikimedia.org/wikipedia/commons/a/a1/Bronze_Star_ribbon.svg *License:* Public domain *Contributors:* Vectorized from raster image <a href='//commons.wikimedia.org/wiki/File:Bronzestar_rib.png' class='image'><img alt='Bronzestar_rib.png' src='https://upload.wikimedia.org/wikipedia/commons/thumb/8/8b/Bronzestar_rib.png/100px-Bronzestar_rib.png' width='100' height='28' srcset='https://upload.wikimedia.org/wikipedia/commons/8/8b/Bronzestar_rib.png 1.5x, https://upload.wikimedia.org/wikipedia/commons/8/8b/Bronzestar_rib.png 2x' data-file-width='106' data-file-height='30' /></a> *Original artist:* Ipankonin

- **File:CNS_Kunming_(DDG-172).jpg** *Source:* https://upload.wikimedia.org/wikipedia/commons/3/3f/CNS_Kunming_%28DDG-172%29. jpg *License:* CC BY-SA 4.0 *Contributors:* Own work *Original artist:* 樊永强

- **File:Cheminee_tribord_du_forbin.JPG** *Source:* https://upload.wikimedia.org/wikipedia/commons/3/3d/Cheminee_tribord_du_forbin.JPG *License:* CC BY-SA 3.0 *Contributors:* Own work *Original artist:* Bublegun

- **File:Combat_Distinguishing_Device.png** *Source:* https://upload.wikimedia.org/wikipedia/commons/3/37/Combat_Distinguishing_Device. png *License:* CC BY-SA 3.0 *Contributors:* Own work *Original artist:* EricSerge

- **File:Commons-logo.svg** *Source:* https://upload.wikimedia.org/wikipedia/en/4/4a/Commons-logo.svg *License:* CC-BY-SA-3.0 *Contributors:* ? *Original artist:* ?

- **File:DD(X).png** *Source:* https://upload.wikimedia.org/wikipedia/commons/1/1d/DD%28X%29.png *License:* Public domain *Contributors:* ? *Original artist:* ?

- **File:DD(X)_Advanced_Gun_System.jpg** *Source:* https://upload.wikimedia.org/wikipedia/commons/4/48/DD%28X%29_Advanced_Gun_System.jpg *License:* Public domain *Contributors:* http://www.navweaps.com/Weapons/WNUS_61-62_ags_pics.htm *Original artist:* Unknown<a href='//www.wikidata.org/wiki/Q4233718' title='wikidata:Q4233718'><img alt='wikidata:Q4233718' src='https://upload.wikimedia.org/wikipedia/commons/thumb/f/ff/Wikidata-logo.svg/20px-Wikidata-logo.svg.png' width='20' height='11' srcset='https://upload.wikimedia.org/wikipedia/commons/thumb/f/ff/Wikidata-logo.svg/30px-Wikidata-logo.svg.png 1.5x, https://upload.wikimedia.org/wikipedia/commons/thumb/f/ff/Wikidata-logo.svg/40px-Wikidata-logo.svg.png 2x' data-file-width='1050' data-file-height='590' /></a>

- **File:Defense.gov_News_Photo_000110-N-7495A-007.jpg** *Source:* https://upload.wikimedia.org/wikipedia/commons/5/5b/Defense.gov_News_Photo_000110-N-7495A-007.jpg *License:* Public domain *Contributors:*
  This Image was released by the United States Navy with the ID 000110-N-7495A-007 <a class='external text' href='//commons.wikimedia.org/w/index.php?title=Category:Files_created_by_the_United_States_Navy_with_known_IDs,<span>,&,</span>,filefrom=000110-N-7495A-007#mw-category-media'>(next)</a>.
  This tag does not indicate the copyright status of the attached work. A normal copyright tag is still required. See Commons:Licensing for more information.
  Original artist: Petty Officer 2nd Class Tim Altevogt, U.S. Navy

- **File:Edit-clear.svg** *Source:* https://upload.wikimedia.org/wikipedia/en/f/f2/Edit-clear.svg *License:* Public domain *Contributors:* The *Tango! Desktop Project.* Original artist:
  The people from the Tango! project. And according to the meta-data in the file, specifically: "Andreas Nilsson, and Jakub Steiner (although minimally)."

- **File:Elmo_R._Zumwalt.jpg** *Source:* https://upload.wikimedia.org/wikipedia/commons/b/b3/Elmo_R._Zumwalt.jpg *License:* Public domain *Contributors:*

- Elmo_Zumwalt.jpg *Original artist:* Elmo_Zumwalt.jpg: PHC W. Mason, United States Navy

- **File:Elmo_Zumwalt.jpg** *Source:* https://upload.wikimedia.org/wikipedia/commons/7/7e/Elmo_Zumwalt.jpg *License:* Public domain *Contributors:* http://www.history.navy.mil/photos/images/h97000/h97202kc.htm *Original artist:* PHC W. Mason, United States Navy

- **File:FELIX.jpg** *Source:* https://upload.wikimedia.org/wikipedia/commons/1/14/FELIX.jpg *License:* CC BY-SA 3.0 *Contributors:* Own work *Original artist:* China Crisis

- **File:FEL_principle.png** *Source:* https://upload.wikimedia.org/wikipedia/commons/3/3e/FEL_principle.png *License:* CC-BY-SA-3.0 *Contributors:* Transferred from de.wikipedia to Commons. *Original artist:* Selbst erstellt (Horst Frank)

- **File:F_220_Fregatte_Hamburg_II.jpg** *Source:* https://upload.wikimedia.org/wikipedia/commons/d/dc/F_220_Fregatte_Hamburg_II.jpg *License:* CC-BY-SA-3.0 *Contributors:* taken from German Wikipedia, de:Bild:F 220 Fregatte Hamburg II.jpg *Original artist:* Photographed by de:Benutzer:Soebe

- **File:Jaureguiberry_1915_AWM_J06004.jpeg** *Source:* https://upload.wikimedia.org/wikipedia/commons/6/6a/Jaureguiberry_1915_ AWM_J06004.jpeg *License:* Public domain *Contributors:* This image is available from the Collection Database of the Australian War Memorial under the ID Number: J06004

  This tag does not indicate the copyright status of the attached work. A normal copyright tag is still required. See Commons:Licensing for more information. *Original artist:* photographer not identified

- **File:Koreacloseairsupport1950.JPEG** *Source:* https://upload.wikimedia.org/wikipedia/commons/0/0a/Koreacloseairsupport1950.JPEG *License:* Public domain *Contributors:* U.S. DefenseImagery photo VIRIN: 127-GK-234F-A54388 *Original artist:* Cpl. P. McDonald, USMC

- **File:Korean_Service_Medal_-_Ribbon.svg** *Source:* https://upload.wikimedia.org/wikipedia/commons/1/1b/Korean_Service_Medal_-_ Ribbon.svg *License:* Public domain *Contributors:* Vectorized from raster image <a href='//commons.wikimedia.org/wiki/File:KSMRib.gif' class='image'><img alt='KSMRib.gif' src='https://upload.wikimedia.org/wikipedia/commons/thumb/f/fb/KSMRib.gif/100px-KSMRib.gif' width='100' height='28' srcset='https://upload.wikimedia.org/wikipedia/commons/f/fb/KSMRib.gif 1.5x, https://upload.wikimedia.org/ wikipedia/commons/f/fb/KSMRib.gif 2x' data-file-width='106' data-file-height='30' /></a> From The Institute of Heraldry *Original artist:* Ipankonin

- **File:Legion_of_Merit_ribbon.svg** *Source:* https://upload.wikimedia.org/wikipedia/commons/4/43/Legion_of_Merit_ribbon.svg *License:* Public domain *Contributors:* Vectorized from raster image <a href='//commons.wikimedia.org/wiki/File:Us_legion_of_merit_rib.png' class='image'><img alt='Us legion of merit rib.png' src='https://upload.wikimedia.org/wikipedia/commons/thumb/d/df/Us_legion_of_merit_ rib.png/96px-Us_legion_of_merit_rib.png' width='96' height='26' srcset='https://upload.wikimedia.org/wikipedia/commons/thumb/d/df/Us_ legion_of_merit_rib.png/144px-Us_legion_of_merit_rib.png 1.5x, https://upload.wikimedia.org/wikipedia/commons/thumb/d/df/Us_legion_ of_merit_rib.png/192px-Us_legion_of_merit_rib.png 2x' data-file-width='218' data-file-height='60' /></a> Us legion of merit rib.png *Original artist:* This vector image was created with Inkscape by Ipankonin, and then manually replaced.

- **File:MightyServantRoberts19882turned.jpg** *Source:* https://upload.wikimedia.org/wikipedia/commons/0/0c/ MightyServantRoberts19882turned.jpg *License:* Public domain *Contributors:* digitally edited (turned) by Johann H. Addicks, from U.S. Department of Defense's Defense Visual Information Center image DNST8902238 *Original artist:* Camera operator: Photographer's Mate 2nd Class D. Kevin Elliott

- **File:NBR_3rd_Class_Carriage.jpg** *Source:* https://upload.wikimedia.org/wikipedia/commons/a/a2/NBR_3rd_Class_Carriage.jpg *License:* Public domain *Contributors:* ? *Original artist:* Tongshan at English Wikipedia

- **File:National_Defense_Service_Medal_ribbon.svg** *Source:* https://upload.wikimedia.org/wikipedia/commons/0/0d/National_Defense_ Service_Medal_ribbon.svg *License:* Public domain *Contributors:* Vectorized from raster image <a href='//commons.wikimedia. org/wiki/File:National_Defense_Service_Medal_ribbon.png' class='image'><img alt='National Defense Service Medal ribbon.png' src='https://upload.wikimedia.org/wikipedia/commons/thumb/c/c9/National_Defense_Service_Medal_ribbon.png/100px-National_Defense_ Service_Medal_ribbon.png' width='100' height='28' srcset='https://upload.wikimedia.org/wikipedia/commons/c/c9/National_Defense_ Service_Medal_ribbon.png 1.5x, https://upload.wikimedia.org/wikipedia/commons/c/c9/National_Defense_Service_Medal_ribbon.png 2x' data-file-width='106' data-file-height='30' /></a> *Original artist:* Ipankonin

- **File:Nave_Caio_Duilio.jpg** *Source:* https://upload.wikimedia.org/wikipedia/commons/2/2b/Nave_Caio_Duilio.jpg *License:* CC BY-SA 3.0 *Contributors:* Own work *Original artist:* ItalianLarry

- **File:Navy_Distinguished_Service_ribbon.svg** *Source:* https://upload.wikimedia.org/wikipedia/commons/2/26/Navy_Distinguished_ Service_ribbon.svg *License:* Public domain *Contributors:* Vectorized from raster image <a href='//commons.wikimedia.org/wiki/File: NavDRib.gif' class='image'><img alt='NavDRib.gif' src='https://upload.wikimedia.org/wikipedia/commons/thumb/d/d4/NavDRib. gif/96px-NavDRib.gif' width='96' height='27' srcset='https://upload.wikimedia.org/wikipedia/commons/d/d4/NavDRib.gif 1.5x, https://upload.wikimedia.org/wikipedia/commons/d/d4/NavDRib.gif 2x' data-file-width='106' data-file-height='30' /></a> NavDRib.gif *Original artist:* Ipankonin

- **File:Navy_and_Marine_Corps_Commendation_ribbon.svg** *Source:* https://upload.wikimedia.org/wikipedia/commons/c/cb/Navy_and_ Marine_Corps_Commendation_ribbon.svg *License:* Public domain *Contributors:* ? *Original artist:* ?

- **File:Navyacademylogo.jpg** *Source:* https://upload.wikimedia.org/wikipedia/commons/c/cd/Navyacademylogo.jpg *License:* Public domain *Contributors:* United States Naval Academy: Seal
  http://en.wikipedia.org/wiki/Image:Navyacademylogo.JPG (secondary) *Original artist:* Park Benjamin[#cite_note-USNA-1 [1]]

- **File:Nuvola_apps_kaboodle.svg** *Source:* https://upload.wikimedia.org/wikipedia/commons/1/1b/Nuvola_apps_kaboodle.svg *License:* LGPL *Contributors:* http://ftp.gnome.org/pub/GNOME/sources/gnome-themes-extras/0.9/gnome-themes-extras-0.9.0.tar.gz *Original artist:* David Vignoni / ICON KING

- **File:P_vip.svg** *Source:* https://upload.wikimedia.org/wikipedia/en/6/69/P_vip.svg *License:* PD *Contributors:* ? *Original artist:* ?

- **File:Phliber_rib.png** *Source:* https://upload.wikimedia.org/wikipedia/commons/e/e5/Phliber_rib.png *License:* Public domain *Contributors:* ? *Original artist:* ?

- **File:Question_book-new.svg** *Source:* https://upload.wikimedia.org/wikipedia/en/9/99/Question_book-new.svg *License:* Cc-by-sa-3.0 *Contributors:*
  Created from scratch in Adobe Illustrator. Based on Image:Question book.png created by User:Equazcion *Original artist:* Tkgd2007

- **File:ROKS_Sejong_the_Great_(DDG_991).jpg** *Source:* https://upload.wikimedia.org/wikipedia/commons/5/5d/ROKS_Sejong_the_ Great_%28DDG_991%29.jpg *License:* Public domain *Contributors:* http://www.navy.mil/view_image.asp?id=64981 *Original artist:* U.S. Navy

- **File:RVN_Choung_My_Medal_1st_class_ribbon.png** *Source:* https://upload.wikimedia.org/wikipedia/commons/f/f8/RVN_Choung_My_ Medal_1st_class_ribbon.png *License:* CC BY-SA 4.0 *Contributors:* Own work *Original artist:* EricSerge

This Image was released by the United States Navy with the ID 090825-N-1522S-020 <a class='external text' href='//commons.wikimedia. org/w/index.php?title=Category:Files_created_by_the_United_States_Navy_with_known_IDs,<span>,&,</span>,filefrom=090825-N-1522S-020#mw-category-media'>(next)</a>.

This tag does not indicate the copyright status of the attached work. A normal copyright tag is still required. See Commons:Licensing for more information.

Original artist: U.S. Navy photo by Mass Communication Specialist 1st Class Leah Stiles

- **File:US_Navy_O10_insignia.svg** *Source:* https://upload.wikimedia.org/wikipedia/commons/c/c7/US_Navy_O10_insignia.svg *License:* Public domain *Contributors:* Source: http://www.defenselink.mil/specials/insignias/officers.html *Original artist:* Ipankonin

- **File:US_Navy_O1_insignia.svg** *Source:* https://upload.wikimedia.org/wikipedia/commons/f/f8/US_Navy_O1_insignia.svg *License:* Public domain *Contributors:* Source: http://www.defenselink.mil/specials/insignias/officers.html *Original artist:* Ipankonin

- **File:US_Navy_O2_insignia.svg** *Source:* https://upload.wikimedia.org/wikipedia/commons/3/38/US_Navy_O2_insignia.svg *License:* Public domain *Contributors:* Source: http://www.defenselink.mil/specials/insignias/officers.html *Original artist:* Ipankonin

- **File:US_Navy_O3_insignia.svg** *Source:* https://upload.wikimedia.org/wikipedia/commons/0/02/US_Navy_O3_insignia.svg *License:* Public domain *Contributors:* Source: http://www.defenselink.mil/specials/insignias/officers.html *Original artist:* Ipankonin

- **File:US_Navy_O4_insignia.svg** *Source:* https://upload.wikimedia.org/wikipedia/commons/1/16/US_Navy_O4_insignia.svg *License:* Public domain *Contributors:* Source: http://www.defenselink.mil/specials/insignias/officers.html *Original artist:* Ipankonin

- **File:US_Navy_O5_insignia.svg** *Source:* https://upload.wikimedia.org/wikipedia/commons/b/be/US_Navy_O5_insignia.svg *License:* Public domain *Contributors:* Source: http://www.defenselink.mil/specials/insignias/officers.html *Original artist:* Ipankonin

- **File:US_Navy_O6_insignia.svg** *Source:* https://upload.wikimedia.org/wikipedia/commons/c/c0/US_Navy_O6_insignia.svg *License:* Public domain *Contributors:* Source: http://www.defenselink.mil/specials/insignias/officers.html *Original artist:* Ipankonin

- **File:US_Navy_O7_insignia.svg** *Source:* https://upload.wikimedia.org/wikipedia/commons/d/d9/US_Navy_O7_insignia.svg *License:* Public domain *Contributors:* Source: http://www.defenselink.mil/specials/insignias/officers.html *Original artist:* Ipankonin

- **File:US_Navy_O8_insignia.svg** *Source:* https://upload.wikimedia.org/wikipedia/commons/9/97/US_Navy_O8_insignia.svg *License:* Public domain *Contributors:* Source: http://www.defenselink.mil/specials/insignias/officers.html *Original artist:* Ipankonin

- **File:US_Navy_O9_insignia.svg** *Source:* https://upload.wikimedia.org/wikipedia/commons/0/06/US_Navy_O9_insignia.svg *License:* Public domain *Contributors:* Source: http://www.defenselink.mil/specials/insignias/officers.html *Original artist:* Ipankonin

- **File:US_Navy_Sea_Shadow_stealth_craft.jpg** *Source:* https://upload.wikimedia.org/wikipedia/commons/a/ab/US_Navy_Sea_Shadow_stealth_craft.jpg *License:* Public domain *Contributors:* http://www.chinfo.navy.mil/navpalib/factfile/ships/ship-sea.html *Original artist:* US Navy employee

- **File:Undulator.FELIX.jpg** *Source:* https://upload.wikimedia.org/wikipedia/commons/b/ba/Undulator.FELIX.jpg *License:* CC BY-SA 3.0 *Contributors:* Own work *Original artist:* China Crisis

- **File:United_States_Department_of_the_Navy_Seal.svg** *Source:* https://upload.wikimedia.org/wikipedia/commons/0/09/Seal_of_the_United_States_Department_of_the_Navy.svg *License:* Public domain *Contributors:* Keeleysam *Original artist:* United States Army Institute Of Heraldry

- **File:Uss_Zumwalt.jpg** *Source:* https://upload.wikimedia.org/wikipedia/commons/e/e3/Uss_Zumwalt.jpg *License:* Public domain *Contributors:* http://www.navy.mil/view_image.asp?id=61727 (080723-N-0000X-001) *Original artist:* U.S. Navy photo illustration/Released

- **File:VLS_Caio_Duilio.jpg** *Source:* https://upload.wikimedia.org/wikipedia/commons/d/d4/VLS_Caio_Duilio.jpg *License:* CC BY-SA 3.0 *Contributors:* Own work *Original artist:* Gaetano56

- **File:VLS_MK41_Canister_Types.gif** *Source:* https://upload.wikimedia.org/wikipedia/commons/e/e2/VLS_MK41_Canister_Types.gif *License:* CC BY-SA 3.0 *Contributors:* Собственное произведение *Original artist:* AdmiralHood

- **File:VLS_MK41_Missile_Launch.gif** *Source:* https://upload.wikimedia.org/wikipedia/commons/f/f9/VLS_MK41_Missile_Launch.gif *License:* CC BY-SA 3.0 *Contributors:* Собственное произведение *Original artist:* AdmiralHood

- **File:Vietnam_Service_Ribbon.svg** *Source:* https://upload.wikimedia.org/wikipedia/commons/6/6d/Vietnam_Service_Ribbon.svg *License:* Public domain *Contributors:* Vectorized from raster image 100px *Original artist:* Ipankonin

- **File:Wiktionary-logo-en.svg** *Source:* https://upload.wikimedia.org/wikipedia/commons/f/f8/Wiktionary-logo-en.svg *License:* Public domain *Contributors:* Vector version of Image:Wiktionary-logo-en.png. *Original artist:* Vectorized by Fvasconcellos (talk · contribs), based on original logo tossed together by Brion Vibber

- **File:Wisconsin_museum.JPG** *Source:* https://upload.wikimedia.org/wikipedia/commons/8/82/Wisconsin_museum.JPG *License:* Public domain *Contributors:* Source Specifically *Original artist:* U.S. Navy Photo

- **File:Zumwalt-class_(DDG-1000)_artist'{}s_conception.jpg** *Source:* https://upload.wikimedia.org/wikipedia/commons/3/3d/Zumwalt-class_%28DDG-1000%29_artist%27s_conception.jpg *License:* Public domain *Contributors:* ? *Original artist:* ?

- **File:Zumwalt_Deckplate_Transit.jpg** *Source:* https://upload.wikimedia.org/wikipedia/commons/c/c8/Zumwalt_Deckplate_Transit.jpg *License:* Public domain *Contributors:* https://www.facebook.com/photo.php?fbid=10151102612062823&set=a.10151102611502823.432610.74281347822&type=3&theater *Original artist:* United States Navy

4.3 Content license